THE YORKSHIRE FORAGER

An expert in the flavours, aromas and culinary and
medicinal properties of long-forgotten British plants,
Alysia Vasey works with some of the most accomplished
chefs in the world. She is a forager and supplier of all
things wild. She has been featured in a number of
magazines including *Good Housekeeping*, *Chef* magazine
and *Savour* magazine. More recently she has made
a guest appearance on *Countryfile* and also on
James Martin's *Saturday Morning* cookery show.

'Alysia is at the top of her game, she's always there for any advice or knowledge when we need it'
Paul Leonard, The Forest Side

'Alysia's passion for the foraged world is absolutely infectious.'
Colin McGurran, Winteringham Fields

'Foraging is a unique skill that Alysia has mastered, it takes time and a deep understanding of the land, sea and, most importantly, the seasons.'
Nigel Haworth, Northcote

'What Alysia doesn't know about the hedgerows of the north isn't worth knowing! She has a knowledge of the edible weeds and flowers to feed many palates, be they home cook or Michelin starred.'
Andrew Pern, Star at Harome

'Alysia has the eye for finding the most amazing products'
Lisa Goodwin-Allen,
Northcote

'Alysia's passion and respect for her field really makes our job easy.'
Mark Birchall, Moor Hall

'A true forager.'
Michael Wignall,
The Angel at Hetton

ALYSIA VASEY

THE YORKSHIRE FORAGER

A Wild Food Survival Journey

First published in 2020 by
HEADLINE PUBLISHING GROUP

First published in paperback in 2021 by
HEADLINE PUBLISHING GROUP

1

Cataloguing in Publication Data is available from the British Library

Though it is not its main purpose, this book provides some guidance and tips on safely gathering and using wild food in mainland Britain, but – as the author consistently reminds the reader throughout the book – it is not intended to take the place of comprehensive foraging guides, plenty of which are available for loan or purchase. No one should eat anything taken from the wild unless absolutely certain of what it is, and that it is safe to eat. The author and publishers will accept no liability for any losses or suffering claimed to result from the contents of this book.

Paperback ISBN 978 1 4722 6912 6

Designed and typeset by EM&EN
Printed and bound in Italy by Elcograf S.p.A.

HEADLINE PUBLISHING GROUP
An Hachette UK Company
Carmelite House
50 Victoria Embankment
London EC4Y 0DZ

www.headline.co.uk
www.hachette.co.uk

CONTENTS

PART ONE: A WILD FOOD SURVIVAL JOURNEY 1

One: First Steps 3

Two: Weekend Breaks 12

Three: Branching Out 23

Four: Finding the Path 34

Five: The Seeds are Sown 48

Six: Putting down Roots 66

Seven: Changing Seasons 82

Eight: On the Shoreline 94

Nine: Good Enough to Eat? 110

Ten: Michelin Stars 124

Eleven: Experiments with Flavour 132

Twelve: Adventures in Foraging 143

Thirteen: Changing Nature 159

Fourteen: Yorkshire Foragers 169

Part Two: My Foraging Year 179

Afterword 297

Acknowledgements 299

Index 301

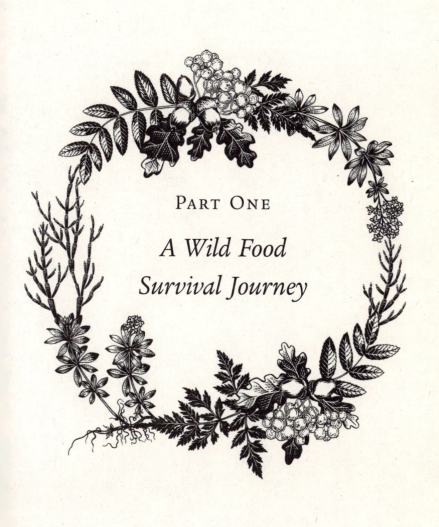

PART ONE

A Wild Food
Survival Journey

FIRST STEPS

'WHAT'S THIS "four-ageing" then?'

'It's called foraging, Grandad,' I said, 'but it's just a trendy word for getting something to eat from the woods, like we used to do.'

Some of my earliest memories are of wandering through the woods and over the moors around my Yorkshire home with Grandad. All through my childhood, my little brother and I spent our weekends following him around the countryside as he searched for plants, nuts, mushrooms and fruit. Depending on the season and the weather, we'd come home with basketfuls of mushrooms, sorrel, young dandelion leaves, wild garlic, bilberries, blackberries, the woodland raspberries that were his favourite, or sweet chestnuts and beechnuts.

I was always more interested than my younger brother Adrian, and though I don't remember Grandad teaching me exactly, I copied him, and by doing what he did, I began to learn just like my mum Barbara had when she'd been a child. Sometimes my mum's brother would come with us too. Robert is only ten years older than me, so he was still a teenager when Adrian and I were first going out for walks with Grandad. My favourite times were going bilberry picking up on Norland Moor above my Nana and Grandad's house.

We'd walk up a green lane between centuries-old stone slabs set on edge as a primitive fencing for cattle pastures. Grandad's border collie, Doogle, would keep bounding around us and plunging into every one of the stone troughs set into the walls, fed by the ice-cold water trickling from the underground springs.

Eventually we emerged on the open moor. It had something of a grim history. The Ladstone Rock there was said to have been used by druids for human sacrifices, and Gallows Pole Hill was named after the public hangings that took place in the bad old days when you could be condemned to death for stealing a loaf of bread.

I'm glad to say small children were spared such tales and I loved being up there. I'd play hide and seek among the boulders and caves, and chalk my name on the Ladstone Rock, though the next shower of rain would always wash it away. When my little legs got tired, Grandad would hoist me up on to his shoulders and I'd ride there as he picked his way through the cotton-grass bogs and strode over the peat and heather in his shirt and tie – Grandad belongs to that generation who always dress quite formally. I've never seen him in a pair of jeans. He'd head for the place where the best bilberries grew, in the old quarries where the stone to build the houses dotting the valley had been dug and blasted out centuries before.

We always knew when the berries were ripe because we'd see flocks of wood pigeons descending to feast on them. We'd clap our hands and a pink and grey cloud would rise suddenly from among the bushes and fly away, carried by the winds. When any of them flew down and landed in the garden, their poo would be bright purple, and if Grandad saw that, we'd be straight up to harvest our share before they scoffed them all. Even when I was quite small, I was happy picking bilberries for hours with my pudgy little fingers. I probably ate as many as I put in the plastic ice-cream tub that Nana had washed out for me. My hands, fingers, lips and face were bright blue by the time we went back home and it would take days for the colour to disappear, but I didn't mind, it was worth it.

Grandad's ninety-five now. He speaks in broad Yorkshire, but with his native Polish accent often breaking through, and he has fine features and thinning hair, which my mum still trims round the sides for him. When he's relaxed he has a serious, almost frowning expression, but there are laughter lines in the corners of his eyes and when he smiles it makes us all smile.

When we got home from the woods Grandad would cook something or make a salad with what we'd found. Back then, Grandad loved to cook, though he wasn't the best chef in the world.

He liked to make everything the day before and then reheat it, so food had often been cooked two or even three times before it reached our plates. Nor had he ever forgotten his Polish roots. I grew up thinking that the accompaniment to the traditional British roast beef Sunday lunch was not Yorkshire puddings, roast potatoes, peas and carrots, but gherkins, pickled cabbage, sauerkraut and maybe a bit of piccalilli as well: fifty per cent British and fifty per cent Polish.

Saturday teatime at Grandad's was closer to a hundred per cent East European: a Polish buffet, with Ukrainian bread and a whole range of salami and kabanos, pickled onions, tomatoes and lettuce, with Polish biscuits to finish. You were always expected to clear your plate, because he hated to see food wasted.

Grandad isn't tall, though he seemed so to me when I was little, because back then he was very strong, with powerful forearms that made light work of digging the heavy clay soil in his garden. His large hands made my tiny ones seem to vanish inside his as we walked along. They felt rough to the touch, callused by a lifetime of manual labour, but they were surprisingly gentle. I remember him stooping to pick up a tiny frog from the grass verge at the edge of the road and carrying it down through the woods. He set it down at the edge of a small pond and we watched as it sat there for a moment, taking in its new surroundings, and then jumped in, disappearing in its muddy depths. We used to see a lot of frogs and newts back then. Sometimes Grandad let us take frogspawn home so we could fill the old sink in the garden and watch the tadpoles develop; it was always fascinating to see their little legs begin to sprout.

When I was old enough to take in more, Grandad would announce, 'Right, I'm going to take you out to the woods today.' He'd stop by a tree and ask, 'Do you know what sort of tree this is, Alysia? And what sort of nuts are they?'

He'd wait for me to take a guess and then put me right. 'Actually, it's a beech tree and these are beechnuts.'

'How do you know?' I'd ask, and he'd launch into a long explanation about how you could tell by the smoothness and colour of the bark, the shape and tint of the leaves and what the nut and its casing looked like.

'And can you eat them, Grandad?'

'Yes, they're tiny little nuts and they're quite soft-shelled, so you can just pick them up, peel them and eat them. They taste good too. Here, try one.'

He was right, they were delicious. In fact I became quite obsessed with them when I was a kid. I passed a group of beech trees on my way to school and in late summer and autumn I'd pause there for ten or fifteen minutes every morning, stuffing myself with all the nuts I could find; the squirrels must have hated me. Picking them off the ground probably wasn't the most health-and-safety-conscious thing to do, because a murder of crows – I love that collective noun for a group of them – used to roost in those trees and covered everything with their droppings. I never caught any terrible diseases though, I just made sure I only ate the clean ones.

Having identified and eaten quite a few, we'd move on.

'Now these are ash trees,' Grandad said. 'The wood is easy to split and it makes good firewood when it's dry, but like most woods, you can't burn it green because it's too full of water, so it spits, steams and crackles and makes too much smoke. These are its seeds, they're called "ash keys" because they look a bit like a key, don't they? And in spring, round about May-time, when they're young and green, you can eat them. They're lovely pickled, which is how we used to have them at my home in Poland when I was a boy. They're not that nice raw but you can eat them if you have to.' They also made great fly whisks: a fat stick with a bunch of ash keys on it was a really good wafter.

As he took me out and about, he gradually taught me the names and uses of everything we came across. I've built on that basic knowledge over the years, but that grounding was all down to Grandad. He has always known everything about those woods and moors, even down to where the best, most berry-laden trees for Christmas decorations grew. When we sat down to our Christmas dinner, we always had holly that was smothered in crimson berries and mistletoe as well, if we had managed to climb up for it, though it was often too high in the trees. I still fill my house with freshly picked holly at Christmas now – I love it in hanging baskets – although there's not as much mistletoe around these days.

Thanks to Grandad, by the time I was seven or eight years old, I could identify every tree, plant, flower, animal and bird that we came across. I knew which plants, nuts, berries and mushrooms were safe to eat and which had to be left alone. I don't think it ever occurred to me at the time that there was anything unusual in such a young child being able to do that; I think I just assumed that everyone knew that stuff.

It was only later that I discovered how exceptional it was. None of my school friends knew the names of anything and, as I found out one autumn day, my teacher at primary school was equally in the dark. It was conker season and our teacher had organised a competition to see who could find the biggest and best one, so she took us out to collect them from the woods at the bottom of the hill, by the river. When we got there, she pointed to a tree and said, 'OK children, see who can find the biggest conker.'

At once the others started rummaging through the fallen leaves or throwing sticks up into the branches, trying to dislodge the nuts, but I stared at the tree for a moment and then put my hand up. 'Please, Miss,' I said. 'That's not a conker tree.'

'Yes it is, Alysia.'

'No Miss, honestly, it isn't.'

'Yes it is.'

Had I been a little older and wiser I might have left it there, since embarrassing a teacher in front of the class is never a good idea, but I was young, innocent and head-strong, and I knew I was right. 'Miss, my grandad knows everything about trees and he told me it's a walnut tree, not a conker tree.'

Although conker, sweet chestnut and walnut casings are all much the same size, Grandad had taught me that they're easy enough to tell apart. Conkers have slightly knobbly casings while sweet chestnuts have a spiky outer skin and walnuts are really smooth. You definitely need to know which are which, because inside, sweet chestnuts and conkers look quite similar, except that chestnuts have a 'tail' on them and there's a double nut inside when you open them. The distinction is important, because while you can eat sweet chestnuts – they're delicious, with a warm, earthy taste that to me is a little like a sweet potato – horse chestnuts, what we call conkers, are poisonous. Strangely enough, you mustn't eat the flowers of sweet chestnut trees, because they are poisonous. Don't ask me how a poisonous flower can produce an edible nut, but it does; it's just one of the tricks that Nature plays.

My teacher was rapidly losing patience with me. 'I'm telling you they're conkers, Alysia.'

'No, Miss, honestly, they're walnuts.'

'No, they're conkers. Here, I'll show you . . .'

She picked one up and dug her nails into the outer skin. After a few seconds' work she had exposed the nut inside. There were a few beats of silence while she looked at it and then, avoiding my eye, she called out, 'These are walnuts, children. Now, who can find me a conker tree?'

I could have done that easily, but the look she directed at me suggested it would be better to keep quiet and let someone else find it. I was walking through that wood recently, almost forty years later. The tree is still there, and they're still definitely walnuts.

One time when Grandad, Adrian and I were out in the woods, Grandad pointed to the trunk of a tree and said, 'Look closely. Do you notice anything?'

I couldn't see anything unusual about it until he showed us the outline of a moth, so beautifully camouflaged that it was virtually invisible against the bark.

'Nature's like that,' he said. 'The harder you look, the more you see.'

He would also show us some of the more intriguing natural curiosities like the mushroom – a lurid boletus – that he picked one day. It had a reddish yellow cap and looked perfectly ordinary until he drew his fingernail across the top of it, whereupon it instantly went a vivid electric blue colour – the same shade as Lady Diana's eyeliner. At the time Diana was all over the magazines and papers looking very glamorous, and I couldn't wait to get into my teens so I'd be allowed to buy an electric blue eyeliner of my own. Now I know that the change of colour is caused by a chemical reaction, as compounds in the mushroom are oxidised when its cell walls are broken and exposed to the air, but back then it just seemed that, on top of all his other talents, Grandad was a magician. He used to bend his right thumb and put his left one on top of it, making it look like he'd broken it in half. He was good at doing things like that to make us laugh.

*

I CAN REMEMBER asking Grandad, 'How do you know all this?' but all he said in reply was, 'I learned it when I was young.' He fell silent then, staring at the ground, though it felt as if it was not the earth at his feet that he was seeing. I waited, feeling anxious without really knowing why, while he remained motionless, his expression clouded. Eventually I tugged at his sleeve. 'Grandad, are you all right?'

I shyly slipped my hand into his and he gave it a squeeze, and when he looked down at me again, the familiar twinkle was back in his eyes. 'Now let's find something to eat, shall we? If you learn what I learned when I was young, you'll never go hungry, because if you know where to look and what to look for, Nature will always take care of you. Here, let me show you.' He gave me a broad smile and then we were off again, deeper into the woods and up on to the moors, with Grandad pointing out the edible plants – and the poisonous ones – as we walked along.

Many years passed and I was an adult myself before I discovered what he had been thinking about that day, and how he had learned that Nature would provide.

CHAPTER TWO

WEEKEND BREAKS

THE PLACE WHERE I grew up is the Ryburn Valley, a narrow, steep-sided offshoot of Calderdale, one of the heartlands of the Yorkshire wool trade. The valley's flanks are covered with deciduous woods and above them is wide-open heather and bilberry moorland – the 'tops' as we call them in Yorkshire – studded with peat bogs and marshes.

My dad had joined the police as a cadet straight from school and risen through the ranks to become a sergeant. He was the local bobby in the pretty village of Ripponden and we lived in a house on the edge of the village with the police station attached. It had an outside phone for emergency use and in those days, when there were no mobile phones, on many a Saturday afternoon the lads from the football pitch ran across the road and used it to summon an ambulance for the latest team-mate to have broken his leg. When we were little, Mum was at home looking after me and Adrian, but before she had us, she had worked in a carpet factory designing colours for the patterns. Ours was navy with multi-coloured spirals – she was really proud of that.

In early 1979, our idyllic childhood came to an abrupt halt. Dad had a heart attack one Friday night. My mum used the station phone to call an ambulance, but by the time it arrived, the village doctor, who lived just a few doors up, had already pronounced him dead. I was seven years old and Adrian only six. Dad was just thirty-one. He drank endless cups of coffee, mixing it with water from the hot tap, and smoked a couple of packs of cigarettes a day, but was a really fit man. However, he had always told my mum that he'd die young, because his parents and a few other members of his family had done so, and sadly he wasn't wrong.

I remember that day only too well. Like most weekends, we had been dropped off at Nana and Grandad's. Mum and Dad usually had a night out on Fridays, so Mum got ready while Dad did the honours and drove us there – he had a new car, so he took every opportunity to be out in it. Excited to be going to Nana and Grandad's, we kissed him goodbye, not knowing that it was the last time we would ever see him.

The next morning, to keep me occupied, Nana had given me a shoe box, a Littlewoods catalogue, some white glue and a pair of scissors, and asked me to make a living room out of it for her. I was in my element, spending ages cutting pictures of standard lamps and sofas out of the catalogue and pasting them on to the sides of the box, complete with carpets, rugs, curtains and furniture. Then Mum arrived, tear-stained and ashen-faced, and my world changed forever. I can still remember Adrian sitting on the floor, surrounded by bits of his train set, trying to understand what had just happened.

APART FROM the trauma of losing him, my dad's death left us struggling to get by. He died just after the 'Winter of Discontent', when almost every group of public sector workers in Britain had been on strike. It had been a winter of electricity rationing, and I remember Mum keeping lots of candles in the cupboards and us having to get dressed under the covers because there was no heating. Sometimes Mum even put the gas oven on in the kitchen to keep the room warm. The strikes paved the way for Margaret Thatcher to become Prime Minister on a pledge to rein in the power of the unions, and one of the first things she did was to give the police a huge pay rise – she knew she'd need them on her side if she was going to take on the unions and especially the miners – but my dad had died a couple of weeks before it came through, so my mum was left with only a small widow's pension. We were so short of money that there was no question of us ever having elaborate toys, doll's houses or bikes; the woods were our playroom and we improvised most of our toys and games from what we found in them, but we weren't unhappy, you don't miss what you don't have.

We were surrounded by open country, with a wood behind the house and another across the road. The one behind us was less thickly planted, with open fields among the clumps of trees, but I'd guess that it was much older than the other one, because there were more huge trees and more goosegrass among them, usually a sign of ancient woodland. There were oaks, beeches and birches growing there, and although birch trees are not particularly long-lived, there were really big beeches and oaks that must have been there hundreds of years, and some huge yews that might well have been even older.

Just along the road – a very steep, cobbled track – a dark tunnel led beneath the canopy. Dating from the days when goods were carried on packhorses, it was so narrow that you would have struggled to get a cart down it, let alone a car. At the bottom you could wade across the river at the old ford, or cross it on a little iron footbridge. A bit further on there was another bridge over the old railway line that used to run right up the valley. Its closure long pre-dated the infamous Dr Beeching's cuts of the 1960s, because it shut to passenger traffic way back in 1929. Although my mum can remember waving at the trains going past when she was a little girl, it closed to goods traffic as well in the late 1950s and Nature soon re-colonised it. Leaf mould covered the ballast and trees self-seeded at the edges of the track, but enough local people walked along it to maintain a path.

With Adrian in tow, I could walk through the woods along the old railway line all the way to Nana and Grandad's house three miles away in Norland. Instead of the boring road with cars and lorries whizzing past, which felt like it took forever to get there, we were deep in among the trees and it was never boring in there; we always found something to look at and do. It would seem like we'd only been in there for five minutes when it might have been an hour or more.

Much of the line ran in a cutting – you could still see the marks of the navvies' chisels and drills on the sandstone bedrock they had cut and blasted their way through – but there were other places where it emerged into the open and there were views through the trees out across the valley. A series of bridges over the line, built from beautifully dressed stone with elegant, wrought iron balustrades, led to Victorian houses deep in the woods. We gave each bridge a number so at any point we always knew how far we were from our house and from Nana and Grandad's.

My brother and I were pretty much left to our own devices and I relished our independence. We played outdoors all day, every day, but then all the kids did back then and every trip into the woods was an adventure. Once, Adrian found a half-buried medieval sword complete with a rotted leather scabbard. I can still remember our excitement at finding this extraordinary prize. We played with it all afternoon, imagining ourselves as knights of the Round Table battling enemies and rescuing damsels in distress – I was always a knight, never a damsel.

We eventually got bored with it, of course, and just dropped it and wandered off. The chances are it's now in a museum somewhere, but I'd love to go back and try to find it. I might also come across the three bottles of vodka left over from a family party many years later, that Adrian and I stole when we were teenagers and hid under a tree somewhere. We never found them again either; I hope that someone did and had a great party with them.

Quite understandably, my mum had her hands full trying to cope with her grief at my father's death and earning enough money to keep her young family, so my brother and I had to become much more self-reliant. Fortunately Mum had already taught me to cook. When she and Dad first got married, Mum's cooking was not the best and Dad encouraged her to go to catering college. He might have regretted it when he found himself eating puff pastry horns stuffed with grapes and chicken in a wine sauce, and various bizarre desserts, but she became a really good cook. She was really proud that she was able to pass this on to my brother and me, when she'd learned something at college she'd always show us how to do it when she got home. So for a while after Dad's death, I did a lot of the cooking, though at first we pretty much had to live on ratatouille because that was one of the only things I could do; I couldn't pronounce it, but I could make it.

I managed to learn how to make chicken liver pâté as well, using *The Police Wives' Recipe Book* that had been given to Mum when she got married. She had also been presented with a knitting pattern for a pullover that, like all police wives then, she was expected to knit for her husband. Graciously, the police gave the lucky wives the right coloured wool and although she'd never knitted in her life, Mum gave it a good go, with guidance from Nana who was an expert knitter and always had a baby outfit on the go, whether there was a baby in the family or not.

IN THE AFTERMATH OF our dad's death, the lifesaver for us and probably for Mum too was that Adrian and I went to Nana and Grandad's every weekend. We went there straight from school on Friday afternoon, stayed Friday night, all day Saturday and Sunday morning and then went home again on Sunday afternoon. On Saturday we went shopping in Huddersfield with Nana and chose something for lunch, and then in the afternoon we would sit on the sofa with Grandad watching the wrestling on the TV. He absolutely loved it and genuinely seemed to believe that Big Daddy, Giant Haystacks and the rest of those larger-than-life characters really did hate each other's guts and were actually going at it hammer and tongs, rather than just following a well-choreographed routine. He'd be on the edge of his seat, red-faced with anger and shouting abuse at the telly, although he never swore, but then nobody did apart from Nana, when she was really mad. If you heard her say 'Shit with sugar on it!' you knew to keep well clear till she calmed down again.

As their house was only a small cottage, with the exception of the Saturday afternoon wrestling, whenever the weather was half-decent we'd usually be out of doors, either in Grandad's garden – he grew a lot of the vegetables they ate – in the woods or up on the moors.

My nana and our Aunty Dora, who lived at the other end of their terrace, were always in their kitchens, cooking and baking. Aunty Dora wasn't a real aunty, that's just what we called her. Her father had built the row of three houses in 1905 after a win on the horses. They are tiny, stone-built two-up and two-downs, with just a kitchen and a living room on the ground floor. Dora's father had built one each for his three children, but by the time I was a child there was only Dora still living there. Grandad would give Aunty Dora fruit and things that he had picked or grown in the garden, and she would make them into these delicious chutneys and jams. She had lots of raspberry canes and made the best raspberry jam, even though she was still using a 1930s cooker. Grandad always complained about the pips, but he still ate boatloads of it.

WE LOVED BEING OUT in the woods and would often be gone for hours on end, having our adventures. We didn't care that we were stung by nettles and thistles, chased by swarms of wood wasps and bees, got splinters in our hands from every tree going and were attacked by all sorts of bugs and insects. Although when one afternoon I stepped into mud after watching a John Wayne movie on TV when he'd got stuck in some quicksand, I was so terrified that the mud would do the same thing to me that I abandoned my new wellies and ran for it. Mum had to come back with me, carrying her shovel to dig them out again. Luckily she saw the funny side.

We had endless cuts and scrapes, and were in and out of the A&E department with monotonous regularity, getting stitched up after falling down cliffs or out of trees, but none of it put us off. There was no 'health and safety', nor the helmets, knee- and elbow-pads with which kids are cosseted now. It was only ever 'How far can you get up that tree?' or 'Bet you can't climb that cliff-face over there', stuff that kids don't seem to do any more.

When he was about fourteen Adrian found an adder warming itself in a patch of sunlight. When he spotted it, he grabbed hold of its tail, whereupon it turned round and bit him on the knuckle – the only truly poisonous snake in Britain, and he'd managed to get himself bitten by one. His arm swelled up like a balloon and he was in intensive care for a few days until his body had worked the poison out of his system and his arm had deflated enough for him to be allowed home. There was no bedside vigil, we all just left him to it until he was well enough to come home again; it never crossed our minds for a second that he wouldn't.

He was pretty accident-prone. He was playing 'throw the javelin at each other' in the garden one day with his friend, who managed to skewer Adrian's foot to the ground. I remember Adrian coming crying into the kitchen, but his only concern was that Mum would 'do him' for putting a hole in his brand new trainers.

I often used to make him crouch behind a wheelbarrow at the bottom of the garden wearing a motorcycle helmet, so that I could shoot empty baked-bean tins off his head with an air rifle. I'm not sure what he got out of it but the main attraction for me was that he could replace the tins himself, ready for the next shot, and save me having to walk down the garden to do so. Luckily I was a very good shot, so I never hit him and it was tremendous fun, though it brings me out in a cold sweat to think about it now.

We ate a pretty healthy diet, with lots of home-grown veg and plenty of wild food too. We got covered in muck and must have swallowed plenty of it as well, so our immune systems were probably supercharged, whereas kids today, who live in environments where every surface has been wiped down with an anti-bacterial cleaner, seem to be vulnerable to every passing bug, allergen and food intolerance. We got the typical childhood diseases: measles, chicken pox, mumps and glandular fever – all the nasty stuff – but I don't remember us ever having colds or flu, or stomach bugs, probably because we were out in the fresh air all the time. I inconveniently got chicken pox when we were on our annual holiday, staying with my Aunty Marilyn at her house in Blackpool, but it was no bother. Mum simply got a paintbrush out, painted me from head to toe with pink calamine lotion, then tied my hands to the deckchair on the beach to stop me scratching the spots off – very practical is my mum.

WHETHER WE WERE near Nana and Grandad's or at home in Ripponden, we could be found damming streams, climbing our favourite trees – horse chestnuts because they had branches that hung low enough to the ground for us to reach – and building dens in the undergrowth or among the rocks. Then we'd hide inside them and eat the jam sandwiches we'd brought with us. We'd made them ourselves and weren't really supposed to have them, but things always seemed to taste better when they weren't allowed.

Practically everything in our pantry had been hollowed out. Jars of jam looked full but we had tunnelled down the middle of them; Mum's giant tubs of glacé cherries were only a single layer thick on the outside of the tub, because we had eaten all the ones from the middle. The same went for the lemon curd and orange marmalade.

When we weren't stuffing down jam sandwiches, we were trying some of the things that Grandad had shown us were good to eat. On a sunny spring day there was nothing nicer than to perch on a tussock of the goosegrass that thrived on the steep slopes and boulder clay of the woods; it looked wiry but was soft to the touch, and even when the ground was wet, the raised humps it formed gave a dry place to sit. It was also good to eat. As the name suggests, geese eat it, and when they need to eat grass to help them digest their food, pets and especially dogs always choose goosegrass over anything else, suggesting that it must have good digestive properties. I loved it too; I'd pull one of the long, dark green stems from its sheath and nibble on the end. The sweet, tender, new growth seemed like – it was – the taste of spring.

Sometimes, when I was out early enough, it was still a bit foggy, the kind where it hovers just above the ground but can still make all your clothes a bit damp. Fog always meant wellies, as the dewdrops saturated the grass and normal shoes would be swamped. I loved the spring when you could smell the bluebells as you walked through them. The bluebell stems slipped easily from their sheaths and you were allowed to pick them in those days, so we would go home with an armful. They looked fabulous for about three hours but then drooped, as all hollow-stemmed flowers do. The tops might have looked barren and deserted to a casual glance, but in spring and summer they were alive with movement and sound. The white tufts of cotton-grass waving in the breeze showed us where the ground was boggy and we could pick a drier course among the tussocks of heather and peat. Red grouse were so beautifully camouflaged as they crouched in the heather that you could almost step on them before they burst out of cover, flapping away in front of us. They skimmed away, hugging the contours of the ground in the flight that earned them their local nickname of 'low flyers', before settling back among the heather, giving their harsh call, which sounds to me like the gobbling of a very distressed turkey. The one-note peeping of the golden plovers on their nests was almost as strange a sound, and was always eerie when the cloud came down, shrouding the tops.

In spring I would often see hares 'boxing' as they battled for mates. It used to be thought that the combatants were males battling for supremacy, but recent research has revealed that one is always a female, testing the male's strength and fighting skills before deciding whether to mate with him.

The moors were covered in wild bell heather, with deep purple flowers and the odd patch of white. If I walked through it in late summer, when the heather was in bloom, clouds of pollen would rise from it, so dense it was like a mist, but one carrying a beautiful, intense scent.

I never thought about it then. I took it all for granted; it was part of our lives. Only now, as I wander through those very same places, do I take the time to reflect on how lucky I was to grow up surrounded by such beauty.

CHAPTER THREE

BRANCHING OUT

THE DEATH OF MY FATHER was the first of a series of tragedies that shaped the way I saw the world. In rapid succession, my Aunty Lorraine, aged just twenty-four, died in a car accident, one of my best friends was knocked off his bicycle and killed, and then my first boyfriend had a terminal brain haemorrhage after being hit on the head by a softball. My friend Glen bought a motor scooter to do his paper round, but didn't load it evenly. As he took a bend, the unstable scooter skidded off the road and fatally crashed down the steep hillside. Four of my friends then drowned in a freak accident when they went surfing and were sucked under by a giant wave and swept out to sea by the undertow. Being made so aware of death at such a young age made me appreciate life. I wanted to pack as much stuff in as possible, and I believed I could.

No one I knew – apart from Nana and Grandad – seemed to live long, making me believe that I had to get out there and do everything I wanted to do fast, because I was not likely to have much time on this planet, but it took me an age to discover my true vocation.

So, when I was still only thirteen, I lied about my age to a local café owner and got a part-time job as a waitress to help Mum with the family finances. By then I had already tried and failed to get one at the Borough Market in Halifax; looking back, it probably wasn't a good move to offer the interviewer a sherbet lemon as a bribe. I loved working much more than being at school, and had quite a few waitressing jobs throughout my teenage years.

One of my favourites was in Ripponden's posh, silver service restaurant, Over The Bridge, which was run by a wonderful, exuberant character, Ian Beaumont, and his partner. It closed a while ago – it's now a bed and breakfast – but back then, its regulars included some of the stars of *Coronation Street* and we thought that definitely made Over The Bridge the place to be seen. Unusually for those days, the restaurant had an all-female chef team – head chef, Sue, and Lindsay – who were quite at the cutting edge of food 1980s-style and although the portions were minuscule – it was the era of Nouvelle Cuisine, after all – they were exquisite. It gave me my first experience of fine dining and I fell in love with it.

I can still remember having my first taste of pheasant there. The freshly shot birds had been plucked and hung in the game cellar until they were green. Sue and Lindsay then wiped them down, roasted them in copious amounts of butter and served them with game chips and mangetouts with almonds: delicious. Mangetouts were very new on the scene then, provoking struggles with the pronunciation among some of the local greengrocers. I remember overhearing one of them, a stout, florid-faced man with a very broad Yorkshire accent, trying to explain this curious new vegetable to a sceptical Halifax housewife. 'They're called man-ji towts, love,' he said. 'It's French for "leave nowt".'

Like all teenagers, I was finding my own way. Busy with

work and all the other things teenagers do, there wasn't time for strolling through the woods looking for things to eat with Grandad or going out on the moor picking bilberries. That's not to say I didn't still walk through the woods, I've never stopped doing that, but back then I was more likely to be marching up a hill to walk off my anger or something that had upset me than to look for something to eat.

On my nineteenth birthday I got engaged to a budding pig farmer – it seemed like a good idea at the time – but it proved to be short-lived; I was dumped about four days later. It was probably just as well, given my growing interest in good food, because the meals his mother served and that he wolfed down all seemed to consist of beef mince cuisine.

Nonetheless, getting dumped was a bit of a shock to the system, and on the spur of the moment, partly in search of adventure and a chance to see the world, but mainly just to get away from Yorkshire to somewhere – anywhere – else, I went and joined the Royal Navy. I was not exactly over-qualified. Partly as a result of nicking off from lessons so often, I'd left school with a single GCE – in Art – though the school wasn't the best either and, being the 1980s, there were more strikes than lessons. So I had to sit an aptitude test at the Royal Naval Recruitment Centre in Leeds before they'd accept me as a recruit. I always knew I wasn't stupid but when I scored ninety-six per cent on the test, I felt vindicated. As a result, I would be given a choice of jobs once I'd completed basic training.

In February 1991 I arrived at Plymouth train station – more than 300 miles from home – wearing jeans and a bright green Benetton jumper with an appliqued parrot on the front, having signed on for twenty-two years. Basic training was held at HMS *Raleigh* at Torpoint, a sprawling base covering several square miles, which was only accessible by a small ferry across the river. Looking back, I believe this was on purpose, like the old days at the navy hospital – where sailors could get better, but couldn't escape.

I didn't particularly enjoy being there. Not many enjoy basic training, but then I suppose it's not meant to be enjoyable. However, I'm not a quitter, so with gritted teeth, I completed my six weeks. Then of course I had a choice, and I fancied trying Cornwall, somewhere I'd never been. So after the training I was assigned to the Navy's Met Office at RNAS Culdrose down at Helston on the Lizard peninsula, which was like Butlin's in comparison.

The landscape was a complete contrast to home and I went walking all the time down to the Loe Bar on the coast, or cycled across to Porthleven. Cornwall's climate can be sub-tropical and I was intrigued by the plants, so many of which – like wild gladioli – I had never seen before. But within a few weeks, bored by the work and unimpressed by the short skirts, blue eyeshadow and pink lipstick that seemed to be de rigueur for all the Navy's 'weather girls', I requested a transfer and became an apprentice helicopter engineer. I travelled to HMS *Daedalus* in Gosport, Hampshire, for another helping of basic training, which we did on Wessex helicopters, even though we were going to be working on Sea Kings as soon as we graduated. Apart from the rotor blades, literally every single thing on a Wessex is in a different place from a Sea King, so it was a bit like learning how to put a motorbike together and then being let loose on a tank, but it must have made sense to someone.

Having completed the training, I was back to RNAS Culdrose to work as an engineer on Sea Kings, in that era still a very unusual role for a woman. I served for almost five years at Culdrose and I loved being at sea and travelling the world. I got my medal for the UN peacekeeping force in Bosnia, though in truth we were a giant offshore airport, providing vital reconnaissance missions and medical evacuation, so we didn't see much of the action.

In time I moved away from the base to a small village called Pendeen on the St Just peninsula, just a few miles from Land's End. It was glorious. From my bedroom window I could see cargo ships loaded with shipping containers making their way out through the south-west approaches to the Atlantic Ocean, heading for ports all over the world. On a bright clear day I could sit on the wall in my garden and if the notorious Pendeen fog wasn't in full swing, I could make out the Scilly Isles about thirty miles away. The local lighthouse had a wondrous foghorn which seemed to sound daily throughout autumn, winter and spring. It was eventually replaced by an electronic squealing version. The original one was much better but I think the locals have got used to the new one now.

Pendeen was where I learned about all the wonderful things there are to eat near the seashore, and where I fell in love with coastal foraging. I got to know George Cass, the landlady of the local pub, the Radjel Inn ('radjel' is Cornish for a fox's den or lair) and we walked our dogs together. I had a German shepherd called Padstow. When I got him I'd been planning to get a bitch and call her Demelza, which is the name of a village in Cornwall, but I'd ended up getting a male. On the way home from collecting him, we passed

Demelza and then Padstow, the coastal town in north Cornwall now best known as the home of seafood chef Rick Stein. So my German shepherd puppy became Padstow. The name did prove a bit awkward once when he ran off in Padstow and I was running round the town calling his name. Eventually I got a tip-off and found him in a pub being fed pork scratchings.

George was very knowledgeable about plants and we would often walk parts of the coastline together having a good look at everything. Because Cornwall is a largely coastal county, with a very different terrain, bedrock, soil and climate from Yorkshire, I found that it was full of plants I wasn't that familiar with. There is always someone else who knows more, and constant learning is the key thing with foraging. You get a good level of knowledge but you're always aware that however much you learn there is still a lot more that you don't know about. George was very good at identifying the plants and nearly all of them were new to me, so I listened and learned. Pendeen had been a tin mining village, and mining had long ago stripped all the trees from the moors, leaving a lot of heathlands, so there wasn't much to forage up on the tops, but when we got down into the woods, and valleys, and along the coast, it was a different climate – quite subtropical – and there were lots of different plants to find and forage.

Portheras Cove was about a mile away down a long lane with a row of terraced houses at the end, painted white and reflecting the midday sun. When you got down to the cove there were three beaches, one where the tourists went and two others which were a bit trickier to get to but when you did you more or less had them to yourself, and the cliffs were full of sea anemones, thrift, rock samphire and pennywort.

Loads of pennywort grew out of the cliffs and granite walls surrounding us and my garden was full from top to toe with the stuff, which was just as well, because I ate it by the ton. It's a wonderful plant with rounded, succulent leaves that look a bit like nasturtium leaves and taste really fresh and lemony, and these big, phallic prongs of purple flowers rising out of them. Pennywort is also a real favourite with chefs, but I've never been able to find any in Yorkshire and it turns out that they only really thrive in Cornwall, a bit of Devon and a handful of other coastal areas. The gentler climate there may well have something to do with it, but I suspect that there is something else, perhaps some mineral in the granite bedrock that they thrive on, and if that's not there, they don't grow well. You do get floating pennywort elsewhere, but that's not the same stuff at all – it's on the DEFRA list as a 'nuisance species', because it grows on the surface of rivers and ponds and spreads so fast that it chokes them of oxygen.

I met some real characters in Cornwall. There was a guy who lived in a hut on the beach. He did a bit of fishing in summer and doubled up as the local coalman, and whenever he'd finished his coal round, he used to wash in the granite trough fed by water from the village well – it was big enough to sit in and have a proper bath. You'd see him sitting there, stark naked and quite unselfconscious about it, washing himself down to get the coal dust off. No one thought anything about it, you'd just stroll along, and call out, 'Morning!' as you passed by. I'd often come down with Padstow and pass the time of day with him and I usually either bought or bartered fresh fish from him. Sometimes it was a bit tricky walking back up with a very enthusiastic German shepherd on a lead in one hand and some equally enthusiastic live mackerel in a carrier bag in the other.

After I had been in the Navy a few years, Mum had moved down to Cornwall too. Like me, she had fancied a change of scene and had moved to Four Lanes, a village south of Redruth, which is a bit further east than Pendeen. One of my favourite memories of mine and Mum's time in the south-west was a night when we got the chance to join in with a traditional Cornish pastime. There had been a big storm and a boat had been wrecked on the nearby cliffs. The word went round Pendeen like wildfire: stuff was being washed up on the beach. It was pure *Poldark*, or that old Ealing comedy *Whisky Galore*, except that sadly there was no whisky involved. The ship had been carrying a mixed cargo, and each bay along seemed to be getting a different consignment, depending on how far the various packages had floated. Newquay was treated to a load of kids' trainers; the next beach along got industrial rolls of clingfilm, and we got bales of tobacco, so we probably had the best of it.

When I heard the news I was straight on the phone to my mum: 'Mum, you've got to get down here quick; we need to go wrecking together!'

Mum is always up for an adventure and came racing over. We got the biggest rucksacks we had – massive, sixty-six litre ones – and walked through the blue fog that was hanging over the cliffs. We could hardly see our hands in front of our faces, but the air was heavy with the smell of tobacco. As we went down the path, we met a steady stream of people coming the other way, all staggering under the weight of their spoils in every kind of bag and container they could find.

When we got down to the beach, the entire population

of the village seemed to be there, but there was plenty for everybody. The sand was covered with big white bales of pure tobacco leaf, and we could see masses more bobbing about on the incoming tide. There was so much of it that they wound up using most of it as mulch on their gardens. I bet the village shop didn't sell a single cigarette for the next decade.

The bales were so tightly wrapped we couldn't get into them. In the end, in desperation, Mum gave Padstow and her German shepherd, Tiggy, the job. They were used to digging up roots and things in the garden, and made short work of it. As a show of community spirit, we then lent the dogs to some of the other villagers while we packed our rucksacks with as much as they would hold. Then we trudged back up the cliff path with them on our backs.

It was only when we got back to the house that we started to think about what we were going to do with it all. We didn't even smoke. Following the advice of locals who could remember similar hauls in the past, we put it in the bath in batches and soaked it with some cheap port, to remove any possibility of sea tang, if salt water had somehow penetrated the wrappings. We then dried and packed it off to my brother, who went round all the pubs and clubs in Halifax and flogged it as rolling tobacco. You can't keep a good forager down.

NOT LONG AFTER Mum had moved down to Cornwall, I decided it was time to pick up where I had left off with my education, so I left the Navy and enrolled for a foundation year at Camborne College. As Four Lanes was closer to college I moved in with Mum, which she was really pleased about as she'd been living on her own.

Having completed the foundation year, I secured a place at the University of East Anglia, studying law and politics, with the inclination and hope of becoming a barrister. Mum was really disappointed to see me move away, but I knew by then that Cornwall was a place to retire, not one to build a life. I was ready for a whole new chapter in a whole new place.

East Anglia provided me with another new landscape, and even while I was at uni, I would still collect wild food if I happened to see it while I was out. I could go down by the lake in the middle of the campus and pick a couple of varieties of wild mushrooms that I was confident I could identify. My friends there used to think that was amazing, and even more so when they discovered I could cook them too. What with Mum passing on her love of cooking and my days waitressing in posh restaurants, I'd become a bit of a foodie, so I was able to knock up some pretty spectacular dishes.

In my final year at uni I answered an advert for a vacation job at cruise company P&O as a port presenter and destination researcher. I had to go around researching destinations where the cruise passengers could either go on excursions or take part in a whole host of activities, then present them to the passengers in the onboard cinema. Despite the long hours and the need to keep a smile pasted on my face, it seemed like the perfect fit for me as I still loved being at sea. But for some reason P&O kept sending me back to Norway, until it reached the point where I swear I'd sailed into every fjord, seen every reindeer and replica Viking longboat, and ridden on every cable car and steam train in the country.

Towards the end of my degree course I entered a competition run by *The Times* entitled 'Tomorrow's Lawyer', which was a search for future legal talent. To say I was quite pleased when I was chosen to be a finalist would be an understatement. But ironically it ended up leading me in the opposite direction. Going down to London for the prize-giving ceremony was an eye-opener for me – I instantly knew I wasn't going to fit in. I seemed to be in a world full of cloned cardboard cut-outs of people. I didn't even look like any of them, and with my dulcet northern tones, I certainly didn't sound like any of them. It wasn't an inferiority complex, it was just a case of thinking, 'Nope, not for me.'

I toyed with staying on with the P&O job, in the hope of getting sent to somewhere a bit warmer than Norway, but Mum dissuaded me. After I'd gone to university, she'd moved back to Yorkshire with Padstow and Tiggy. Aunty Dora had died and her house was on the market. Meanwhile house prices in Cornwall had rocketed. Mum was able to sell up in Four Lanes and buy Aunty Dora's house for cash. So I left university and moved back to Yorkshire without really thinking about what I was going to do. I'm a positive thinker in that respect, I've always believed that things will have a way of working out.

CHAPTER FOUR

FINDING THE PATH

AT THE AGE OF thirty-four I was really happy to be living with Mum, and next door but one to Nana and Grandad in Norland. Mum and I have a very close relationship. We are each a rock for the other and no matter how much we argue and bicker we know we can rely on each other through thick and thin. In my eyes she is the best mum on the planet.

I did a couple of weeks temping for the Calderdale Youth Offending Team, which turned into about eighteen months. This made me think about training to be a teacher, so to get my foot in the door, I started work as a teaching assistant at a local comprehensive. They gave me all the fourteen- and fifteen-year-olds who had been kicked out of their classes for being disruptive. I guess they thought that if I turned out to be a useless teacher, I wouldn't be doing much damage to kids they'd already written off, but I thought they were great. I also had a secret weapon, a lady named Ang, another teaching assistant and a fount of knowledge. I bugged her constantly about problems I was having and she always had a solution. To this day, long after my teaching stint ended, we have remained very good friends. The kids I taught had a lot more character than most of their contemporaries, and definitely weren't stupid, and I had a cunning survival strategy if they did get awkward. I'd say something like 'You bombastic reprobate!' and completely flummox them.

'You've just insulted me, haven't you?' would be the response. 'What did you just say?'

I'd give an innocent smile and say, 'There's a dictionary, look it up.'

It was teaching by stealth; they seemed to quite enjoy that, and so did I. But I could see clearly from other members of staff that it was no longer the profession it once was and I decided the reward wasn't worth the sacrifice.

Now that I was living with Mum I was able to spend more time with Nana and Grandad than I had in years – and we finally pieced together what lay behind the faraway look in Grandad's eyes when I'd asked him how he knew so much about living off the land.

It wasn't Grandad who told me. We all knew that Grandad was Polish and had moved to Yorkshire after the Second World War, but he had always become visibly upset if I, or anybody else, asked him about his life in Poland, and refused to discuss it. When Mum and I were back living in Yorkshire, she and I joined up the few things that he and Nana had told her, and the fruits of her and her sister's research.

BOGDAN ADAM STEFAN SZPERKA, now known as Dan, was born in Poznań in 1925, and had a younger brother, Tadeusz, known as Ted. Their mother died in childbirth when they were both still young, and the daughter she was carrying was stillborn. Their father, Isydor, a train driver and railway engineer, later remarried and his new wife gave birth to Dan's beloved half-sister, Magdalene, who was ten years younger than him.

Grandad's parents were ethnic Germans. Poland had been partitioned by Prussia and Russia in the early nineteenth century and hundreds of thousands of Germans had settled in western Poland, but after the First World War, the Treaty of Versailles recreated the Polish state. Given the ultimatum of either taking Polish nationality or being expelled, like most other ethnic Germans, his parents opted to become Polish citizens. They were happy at first, but when Hitler's troops invaded Poland in 1939, all those who had changed their nationality were regarded as traitors to the German Reich.

Grandad was lucky to survive the invasion. In the chaos and confusion of the German blitzkrieg – the ferocious bombing and shelling that pulverised the Polish army and devastated many cities – he'd become separated from his father and siblings. As the bombing continued, he was cooped up in an air raid shelter for hours. Eventually, he risked death to go outside and get some fresh air. A couple of minutes later, the shelter took a direct hit, killing everyone inside.

Although he was still only fourteen, my grandad was of working age and he and Ted were both apprenticed to their father. They carried on working on the railways under the new Nazi regime, but Isydor was already under suspicion of passing messages between different parts of the Polish Resistance as he drove through the Nazi-occupied areas. What was left of the Polish army, together with partisans and others on the Nazi death-lists, had retreated to the forests and marshes or gone underground in the cities to wage guerrilla war against the invaders.

They were terrifying times, and Dan came close to being shot one day when he was surrounded by German soldiers who thought he had a gun in his pocket. As one of them held a gun to his head, Dan put his hand in his pocket and slowly took it out again to reveal not a gun but a harmonica. He still had a harmonica when I was a kid and could play it beautifully; I like to think it might have been the same one.

Within weeks of the invasion, Heinrich Himmler, 'Reich Commissioner for the Strengthening of the German Race', began to 'eliminate the influence of such alien parts of the population as constitute a danger to the Reich and the German community'. Tens of thousands of Poles were expelled and tens of thousands more – Jews, communists, socialists, intellectuals, teachers, priests and 'traitors' who had given up their German nationality – were rounded up and sent to forced labour or concentration camps. Their lands, homes and businesses were confiscated and handed to *Volksdeutschen* – German Poles – or the floods of new settlers being brought in to permanently 'Germanise' western Poland. Special courts were set up to try 'terrorists' and savage new laws were introduced. The death penalty was imposed for dodging conscription for forced labour, selling or buying anything on the black market, sheltering Jews or infecting a German soldier with a sexually transmitted disease.

Poznań – now once more given its old German name of Posen – was surrounded by a chain of forts built by the Prussians in the nineteenth century to guard their eastern frontier. One of them, Fort VII, became the first concentration camp in Poland: Konzentrationslager Posen, chosen because it was remote, hidden behind high earth embankments that were smothered in dense vegetation, and surrounded by a wide moat. The bridge across the moat led to a forbidding black tunnel that opened on to a sunken central courtyard. Locals living nearby were expelled and their houses were occupied by camp guards and Gestapo officers, completing Fort VII's isolation from the outside world.

Two or three hundred prisoners were held in each sixteen-by-sixty-foot cell, sleeping on rotten straw or the bare earth. The cells were infested with fleas and rats. In heavy rain the women's cells, which were underground, flooded up to a foot deep. Two typhoid epidemics carried off eighty per cent of the prisoners, though their ultimate fate would have been death anyway; only one man ever escaped from the camp.

Officially branded a prison, Fort VII was actually an extermination camp. As early as October 1939, just one month after the Nazi invasion, 400 patients and staff from mental hospitals around Poznań were killed in a gas chamber in one of the bunkers. Many thousands of Poles and Jews, defined by the Nazis as *Untermenschen* – 'subhuman' – were killed there. Prisoners were beaten, tortured and forced to run up 'the stairway of death' holding heavy stones. When they reached the top, the guards kicked them down again. Up to ten people were shot every day; larger-scale shootings took place outside the fort, with the victims toppled into mass graves.

Isydor, Dan and Ted were forced to man the trains delivering cattle wagons filled with these poor souls to Fort VII, and were well aware of their inevitable fate, because the returning trains were always empty and no one taken there survived more than a few days. The locked wagons had no sanitation and, desperate for water, the prisoners would try to barter for it with whatever valuables they still possessed. Horrified at what they were witnessing, in defiance of the Nazis, Isydor, Dan and Ted risked their lives to give them some. They unlocked the wagons when they could and Isydor slowed the train on the section of line that ran close to the forest so the lucky ones could leap out and try to escape.

It couldn't possibly last, of course, and one day in early 1940, Isydor, Dan and Ted were arrested. With a group of other prisoners, the boys were thrown into the back of a cattle lorry owned and driven by a local farmer, with a German guard riding shotgun alongside him for the journey to Fort VII. Dan and Ted never saw their father again.

Although the worst of the winter was over and the trees were turning green, it was a bitterly cold day and the German guard was shivering in his uniform. He eventually told the farmer to hand over his heavy fur coat. The farmer refused, they argued and fought. As the truck slewed to a halt, Dan, Ted and the others seized their chance, jumped out and ran off.

They split up into small groups to give themselves the best chance of escape, and Dan and Ted kept running, deeper and deeper into the forest, until they thought their lungs would burst. When they heard shooting behind them, they went to ground in a patch of dense undergrowth. They did not come out again until night had fallen and the forest was quiet.

They went hungry that night but began foraging at first light. Before the war, they had often wandered through the forest with their father, picking mushrooms and collecting nuts, berries and edible plants, as many of their fellow-countrymen did, and still do to this day, so they had at least an idea of what they could eat to survive. Water was not a problem; there was no shortage of small streams, lakes and marshes.

They slept in caves or hollows under the trees with leaf

litter or pine needles for bedding. They made fires using dry deadwood broken from the trees or windfall branches – where possible, denser woods like juniper or yew – that had been kept clear of the ground by the tangles of undergrowth, so they burned with virtually no smoke. They were constantly alert, their ears attuned to the cries or sudden movements of birds or animals that might signal intruders or German patrols. Sometimes they heard the faint sound of gunshots but it was impossible to tell whether they were fired by hunters or Nazi soldiers.

At first, they could find very little to eat. If it rained and they were lucky, they'd harvest spring mushrooms, such as Jew's ear, from dying elder branches, but it wasn't long before they were racked with hunger.

Desperate for food, they began retracing their escape route but as they neared the edge of the forest, a terrible sight greeted them. Ten corpses, their bodies riddled by bullets and with clouds of flies buzzing around them, were hanging from the branches of a huge oak tree. They recognised many of the faces – the men and boys who had escaped with them, shot and then hung there by the SS and left to rot as a warning to others. Shaken to the core, they hurried back into the heart of the forest.

BEFORE LONG, wild garlic began to grow, and as spring advanced, more mushrooms appeared including the first morels and St George's mushrooms pushing up through the leaf litter. The chicken of the woods – a fungus so named because it tastes a bit like chicken – revealed itself on the branches of oaks and beeches. They collected the nutritious shoots and young cones of pine trees, both had a sweet, citrus taste and were full of vitamin C. They could even strip off the outer bark from the pine trees and eat the resin-filled layer underneath; unlike the shoots and cones, it wasn't particularly pleasant-tasting, but it was what they needed to survive.

The young, green seeds of the ash tree – ash keys – were another vital food. Although, as Grandad would tell me many years later, they don't taste particularly appetising when raw, they're delicious pickled and are now quite a trendy ingredient, but Dan and Ted had to swallow bucket-loads of them that spring, when there was not much else to keep them alive.

The forest covered around fifty square miles and was ancient enough to harbour a rich diversity of plant and animal life. Although they had fled with only what they stood up in and hadn't even a fish hook, like a lot of other things in life, if you absolutely have to do something, you tend to find a way. They had their pocket knives and a talent for improvisation, and they made fish traps out of bendy willow branches and strips of bark, and rabbit snares and bird traps using their bootlaces, rusting wire and discarded tin cans and other objects they found near the paths that they avoided at all other times. With great patience and increasing skill, they trapped anything they could lay their hands on. When it was safe to light one, Dan cooked their food over an open fire – Grandad was always proud that he'd done the cooking – and when it wasn't, they ate it raw.

Remaining deep in the forest, they survived by hunting and foraging for mushrooms, plants, herbs, berries, nuts and fruit, living in perpetual fear of discovery and always alert for danger. From time to time they heard voices or footsteps or saw birds take flight, giving their alarm calls, and at once they hid or moved swiftly away, deeper and deeper into the forest.

The only protection from the Nazi terror was often how fast they could run and how well they could hide.

As the days became months, the Germans never stopped hunting the forest for fugitives, but the search parties were in unfamiliar, difficult terrain and either never searched hard enough or were afraid to go deep enough into the tangled, thorny undergrowth and the swamps and marshes, where their prey might have been waiting in ambush.

Dan and Ted were not the only ones seeking refuge in the forest. Polish troops who had kept their arms and refused to surrender, partisans, deserters, Jews and other runaways from the Nazis were also hiding there, but the boys didn't join forces with them, feeling that their chances of survival were better on their own. That way they lessened the chance of betrayal or being let down by the others' carelessness or the lack of woodcraft that might lead the Nazis to their lair.

Their fears were well justified. Some groups even built elaborate hides, digging underground bunkers with entrances covered by planks to which small branches, twigs and leaves had been painstakingly attached, but remaining in one place like that greatly increased the risk of discovery. Others used green or wet wood for fuel, sending columns of smoke rising above the tree canopy that led to them being found and executed. Others ventured into the villages, or begged, bought or stole food from the farms that fringed the forest, but by doing so they risked betrayal or arrest, or leaving a trail back into the forest that could be followed by Nazi patrols.

Collaborators were another constant threat. They'd come into the forest carrying baskets as if they were just out to gather some mushrooms, but heavily armed German troops would be moving silently through the trees behind them, and a shout was enough to bring them running to arrest, or, more likely, kill anyone they found.

The ones who evaded discovery and death were those, like Dan and Ted, who kept the strictest discipline. The boys changed their campsite every night, sometimes taking it in turns to sleep in outlying barns while the other kept watch, but more often using temporary shelters deep in the forest. One always stood sentry while his brother slept, and they never raised their voices above a murmur, nor showed themselves in the open during daylight. They became survivors.

As time passed, Dan and Ted did sometimes venture out of the forest to steal potatoes or buy food or milk from trusted farmers they had known before the invasion. Even then, they took a circuitous route and one of them would hide on the way back, watching their tracks to see if they were being followed, before linking up with his brother again.

In the winter of 1940, Dan and Ted saw a small group of young men picking their way through the undergrowth. Fearing that they were more collaborators, their first instinct was to remain in hiding, but then they recognised one of them, Tadeus Witkowski, a good friend from their school days. When Dan and Ted came out of their hiding place and spoke to him, they discovered that Tadeus, now operating under the *nom de guerre* of Almor, and his comrades were with the Polish Resistance, the 'Home Army' waging a guerrilla war against the Nazi occupiers.

Dan and Ted decided to join them. They were given some military training by the leader of the group, a Polish army officer who had gone underground after the invasion. They changed their location constantly, making long night marches to new temporary camps, and hid in woods, forests and fields, without the local population even being aware of their presence.

Despite the risks – at first the Nazis murdered twenty, and later a hundred Poles for every German killed or every act of subversion – they carried out ambushes, sabotage and assassinations of Nazi officials, collaborators and informers, using weapons and ammunition dug up from underground stores that had been set up by the Polish army before the war in fear of a German invasion. They derailed trains, dynamited bridges and junctions, and put sugar in the fuel tanks of trucks, causing the engines to seize. Their campaign was so effective that on some railway lines, the Germans were forced to post sentries on every bridge, with two more patrolling every two-hundred-yard section of track in between.

The Resistance was organised in small cells to minimise the risk of betrayal if others were captured, and every man fought under a different name, so that not even their close comrades knew their real names. Dan and Ted spent the rest of the war years with the Żniwiarz – 'Reaper' – Company, eventually fighting in the Żoliborz district of Warsaw, and fired the first shots in the Uprising of 1944. Grandad still has photos of them both in their Uprising uniforms. The battles above and below ground were horrific – the sewers saw some of the most savage fighting – and ninety-two of their comrades were killed. The Reaper Company's role is still commemorated in Warsaw, but the Uprising was eventually crushed. The advancing Soviet armies paused on the outskirts and waited until the Nazis had eliminated the last resistance and razed the city to the ground before attacking and occupying what was left.

*

WHEN THE WAR ended, the new Polish Communist regime expelled hundreds of thousands of ethnic Germans from the country while the regime's Soviet patrons rounded up members of the Home Army. They had waged a ferocious guerrilla war against one invader and the Soviets did not wish to risk them taking up arms against the new occupying power, so tens of thousands of Resistance fighters were arrested, disarmed and either executed or sent to the *gulags*, forced labour camps in the Soviet Union.

Rather than risk the same fate, Dan and Ted evaded capture and left their homeland, making their way west across the border and through the ruins of Germany, foraging for their food along the way, for there was no spare food to be had, with much of the German population close to starvation. They trekked right through the country and further south, eventually reaching a displacement camp in northern Italy. There they were given the choice of emigrating to Australia, America or the UK, and both of them opted for Britain.

When he arrived, Grandad found work in a carpet mill near Bradford in Yorkshire, where he met my nana, Winnie. Nana had also had quite a tough start in life because her mother had died young, and when her father had remarried and had three more children, she and her brother had been passed around various aunties. My Nana ended up going into service and her brother, Jim, joined the army. That was how he met his lovely wife, who was from Hong Kong. He brought her back to Yorkshire with him and they lived near Nana and Grandad. After we'd been out with Grandad we'd always be allowed to go up to theirs and she would give us some reject coconut and cherry cabanas from the Mackintosh's sweet factory in Halifax where she worked.

DAN AND WINNIE married and set up home in the Ryburn Valley, and before long they had four children, including my wonderful mum, Barbara. Grandad kept working in the carpet mills, while Nana brought up the family and worked part-time as a cleaner at the main police station in Sowerby Bridge.

Although Grandad and Nana were never out of work, their wages were low and money was always tight, so to add to the family diet Grandad used the survival skills he had learned in the Polish forests. He almost always worked the night-shift at the carpet mill, as many of the other Polish workers did, and as he made his way home in the early morning light, he'd pick mushrooms, plants, fruit and nuts from the fields, woods and moors and bring them home.

Grandad's half-sister, Magdalene, had survived the war, but she remained in Poland, behind the Iron Curtain. Dan and Ted had managed to stick together like glue throughout the war and the displacement camp, all the way to Britain, but they had not seen her since the day they had fled into the forest. Although they now suspected that their father must have died, they remained unaware of how he had met his end.

Magdalene had tried to tell Grandad what had happened in her letters, but mail to the West was heavily censored and that part of her letters was always blanked out. A photo of their father lying in a coffin did get through, but the tight focus of the photograph made it look like he was lying in a bed, and Grandad hadn't understood the significance of the bandage covering the bullet-hole in Isydor's head.

Grandad remained in ignorance of how his father had died until the late 1980s, when he made his first trip back to Poland since the war and met Magdalene face to face. During the Cold War, Grandad and the other Polish workers at the carpet factory had all been scared to return, but by the late 1980s tensions between the East and West had eased, so they hired a coach and all travelled from the factory to Poland together. It was the first time Grandad had seen his sister in almost fifty years. She told him that after they were arrested by the Nazis and driven away in the cattle truck, the SS had shot Isydor in the head and left him where he fell. Sadly Ted had died from cancer by this time, and so never learned of his father's fate. Magdalene died a few years ago and now lies with Grandad's parents, and his grandparents and great-grandparents, in the old family burial plot in Poznań.

WHEN I'D LEARNED something of Grandad's story from Mum, I did sometimes try to talk to him about it. I might ask him about his childhood in Poland or what it was like in the war. But he would always give me general answers about how he wanted to leave school and become a draughtsman, or about working with his dad on the railways. Grandad has only ever really talked to my nana about that time. He's occasionally talked to my mum and Aunty Marilyn, but not for years now, and I don't think he'll ever talk to any of us about it again.

Chapter Five

THE SEEDS ARE SOWN

I WAS STILL STAYING AT my mum's on my birthday in 2005, when my friends organised a night out for me at the Manchester greyhound stadium. I'd never been to a track before and had no idea how to choose a dog or put on a bet, so I was in the queue at the window, still trying to work out what all the numbers and letters meant, when the man next to me offered to explain. He was tall, dark, and had a twinkle in his eye and a kind face. I was immediately taken with him and was happy to accept his help. He talked me through it all and I placed my bet. By a miracle, it turned out to be a winner – and so did he. Well, that's what I thought at the time, but after fourteen years together I think it's him who won the jackpot.

In 2016 Chris and I got married. I think you get married when you decide you can put up with all of your partner's faults – and it took us a while! It would be Chris's second wedding so a church ceremony wasn't on the cards, and after talking about it a few times we decided to go to Gretna Green. We liked the romance of it all – well, it was more me who was into the romance – and I'd promised him I didn't want a big do. My mum would have to look after our dog while we were away and I said to Chris, 'If Mum can't come, no one else can.' So we had the full white wedding with bouquet, photographer and a reception at a Scottish castle. It was magical – and just for us.

DURING THE FIRST few years Chris and I were together, I was still working with young people. I was increasingly realising that it wasn't for me, but hadn't settled on a different career path to follow. No one has ever launched a career by thinking, 'I know, I'm going to pick weeds for a living,' and nor did I think: 'One day I'm going to start supplying top restaurants with wild food.' But things have a way of finding you if you let them, and it took a set of wasp stings and a truffle hunt to finally launch me as a professional forager.

In the couple of years since I'd returned to Yorkshire, whenever I wasn't at work, I took myself off into the woods behind our old house in Ripponden. The only difference from when I was a kid was that I was now a lot more interested in everything that was going on around me; I was giving the dog a walk but I was keeping an eye out for anything I could gather.

One day I came across an abandoned beehive in a tree that had been blown down in a storm. I only helped myself to a small part of the honey, in case the bees returned; they'd need the rest to survive the winter. The best bit was, I didn't even get stung.

That soon changed on the way home, though, after I trod on the entrance to a wood wasps' nest. I didn't realise what I'd done until they came swarming out, and they were not at all pleased. Tiggy was a very clever German shepherd, and she knew when I said, 'Run!' that I really meant it, but neither of us was up there with Olympic champions in the sprinting stakes. I was always built for comfort, not speed.

After a few seconds, during which we picked up quite a few stings, the wasps disappeared and we started to relax. What I didn't realise was that their angry signals were also being picked up by every other nest in the wood. So every few yards a fresh swarm would appear, barbed bums at the ready. It was quite terrifying; each time I thought we had escaped and let us both come to a gasping halt, another swarm would appear from nowhere. Even when we got through the clanging iron gate at the edge of the wood, Tiggy and I were still being stung to buggery. Those wasps gave both of us a right good going over.

When we got home, Mum was sitting with Nana and Grandad in their garden. As soon as she heard what had happened, Nana was straight down the steps to the house to get her secret Yorkshire sting treatment. Of course you can get perfectly good antihistamine cream in a chemist and that's what you might have expected her to come back out with. But no. In my family you get a bottle of chip-shop malt vinegar squirted at you. And because dogs always come above people in our house, Tiggy was the first one to get a dousing. But, to be fair to Nana, vinegar works, even if it does make your eyes water and you pong a bit.

I'm prone to stings. I got one in the neck from a bumble bee in the school egg-and-spoon race – please note that I was winning – and several more from a wasp hibernating in a picnic blanket in the middle of winter, when Nana put it on my bed for extra warmth. It was warm alright, warm enough to wake up one very grumpy wasp. But I'd never been stung like this.

Yet strangely enough, as I studied my lumpy, swollen face and arms in our bathroom mirror, the only thing that really consumed me was how the wasps had communicated with each other. I started to think about how it's not just insects, but wild birds, animals and plants all connect and communicate with each other. I started to wonder why some plants can be found in particular places but not in others, which often appear identical.

From then on, whenever I went back into the woods, I became much more observant of everything, not just the flora and fauna, but the bedrock, the soil, the microclimate, the altitude and orientation, and when I was at home, I was constantly on the internet, accumulating vast amounts of information about plants and wildlife I was interested in. And at that point what I had become very interested in were truffles.

Truffles have always been highly valued, both because they taste delicious and because, as a subterranean fungus, they're really tricky to locate. Usually there is no sign of them above ground, but they have a strong smell, apparently very like the pheromones given off by male pigs, which is why sows make such good hunters. It's good for the truffles too, because in order to reproduce, they have to be dug up and eaten by animals, which then spread the spores in their droppings – though when pigs root through the soil with their snouts, they can damage the truffles' delicate mycelia, the network of gossamer-thin filaments through which they get their nutrients, reducing their future yield. As a result, the use of pigs to find truffles has been banned in Italy since 1985. In any case, if you're walking round the woods with a pig on a string, you look a bit weird, and saying you're just taking Percy for a walk may not ring entirely true, especially if you have a small shovel in your other hand. There's also the small matter of foot and mouth disease regulations requiring you to have a licence to move a pig anywhere.

The English summer truffle has a black skin, with rather symmetrical warts on it, that make it look a little like a very large, blackened raspberry. They gravitate towards the surface when they ripen in summer, but you can find them all the way through to autumn, and sometimes as late as November. When cut open, the flesh is cream to pale brown with white marbling early in the season, which darkens later on, and summer truffles are always better later in the year as they mature. Those known as autumn truffles, which actually encompasses about three differing types of similar truffle, are a similar size and colouring, but a darker brown inside and have a stronger taste and smell. One smells of creosote, but then I love the smell of creosote. However, autumn truffles are very uncommon in the British Isles.

The irresistible aroma of a truffle comes from the gas it has gathered from the soil nutrients, which it emits to attract animals who will then dig the truffle up and spread its spores. All the English truffles have a more delicate aroma and taste than continental varieties – in many Italian trains there are signs forbidding the transportation of truffles because they permeate through everything. Some continental truffles are hugely prized like the Italian white Alba truffles and the Perigord black truffles, which are known as 'black diamonds' because the most valuable can sell for over £10,000 a kilo. A single huge white one once sold for £165,000.

Those kinds of prices have led to a bizarre increase in truffle forgery, with Chinese truffles worth a twentieth of the price being passed off as their European cousins. The problem is that a truffle's aroma permeates anything you put in with it, so if you mix a couple of genuine Périgord truffles with a batch of Chinese imitations, they will end up smelling identical, and unless you have a good eye, you may not realise you've been conned until you come to taste them. A really unscrupulous supplier might even try to pass off a knobbly potato full of dirt by hiding it in a basket of genuine truffles. People have been known to fall for this one, but I think you would have to be a bit dim.

SINCE MY OBSESSION BEGAN, I must have bought every truffle book ever published. In those first months, while besieging my local library with requests for ever more obscure books on truffles and devouring every article I could find on the internet, I discovered that the Apennines, where the finest Italian varieties are found, are very similar to the Yorkshire Pennines. Both have the same limestone bedrock, and though the Apennines are much higher – well over 9,000 feet at times – because they are so much further south they have broadly the same climate as my native hills. So I began thinking, 'There must be truffles somewhere in the woods around here.' I discovered that in Roman times they used to dig up truffles, especially English summer truffles, from all over the country and take them back to Rome. They were especially plentiful in the south, where the chalky plains of Wiltshire, East and West Sussex and rolling into Hampshire made for ideal conditions, but there were plenty in the north of England too. So where had they gone since then? I had a pin map of the places they'd been found in the previous two centuries, and finding them became my mission.

There were no short cuts in my search. Although pigs and dogs can detect the scent through layers of soil, it's too subtle for a human nose to do so. I had read that there is a little golden truffle fly that homes in on their scent and hovers above them, so if I were to spot one, I'd know I was in the right place. But despite countless hours with my eyes peeled – eleven o'clock in the morning is peak fly time – I've never seen one yet.

I tried to train our ageing German shepherd, Tiggy, but without a supply of truffles, it was a bit tricky. I did try with truffle oil from the supermarket, but she wasn't really into it and I've learned since then that the flavouring can be synthetic anyway, so I probably never stood a chance. So it was back to the drawing board. I knew that truffles have a symbiotic relationship with the trees they grow beneath and that their mycelia – the umbilical cords of truffles – are like an enormous spider's web. When you see the roots on a tree that's been upturned in a storm, it's nothing compared to the network of mycelia. They can spread through an entire wood, attaching themselves to tree roots and exchanging sugars and nutrients with them; mushrooms can't live without trees, and trees can't live without mushrooms.

The Pennine woods around Norland apparently had all the magic ingredients: the right alkaline soil, a suitable climate and plenty of the beech, oak and birch trees under which truffles grow. The Royal Botanic Gardens at Kew had reports of truffles being found many, many years ago around Bradford and Leeds, which were right on our doorstep. I hoped that since we had the famous 'rhubarb triangle' in this part of Yorkshire, a 'truffle triangle' might not be too far away. So now I was on a mission to prove it.

My principal target was a wood with no mushrooms, only a couple of miles away from one that had a public 'mushroom trail' running through it. This may sound counter-intuitive, but there was a logic to it. Truffles tend to monopolise the soil and its nutrients, leaving virtually nothing for any fungi growing on the surface, so they destroy the competition once they take a hold. I reckoned that when the local conditions were right but there were no mushrooms in sight, it was time to swing into action with my mum's garden dibber. The soil beneath the trees was bare of vegetation and had the scorched-earth look that I was after.

I had come across a reference to an old, long-out-of-print and very rare book, originally published by the Royal Botanic Gardens at Kew, describing every native British truffle. After a series of fruitless searches, I eventually managed to find a copy in a second-hand bookshop and bought it for an eye-watering amount of money. I read it avidly and discovered that, rather than the half-dozen types of British truffle I'd been expecting, there were actually sixty-three different varieties in this country.

So I thought to myself, 'If there are sixty-three, I'm bound to hit one,' and set to work with renewed enthusiasm. When I found a likely-looking tree, I poked a hole with my dibber and then dug down into the soft soil with my fingers, releasing the familiar autumn scents of damp earth and leaf mould. The first few holes I made produced nothing but dirt, but I was eventually rewarded. I made a fresh hole, widened it with my hand, and as I dug down a little deeper, I felt a small, rounded, coarse-skinned shape. I peered down and caught a glimpse of yellow.

'Ooh, this looks a bit good,' I thought and I worked my fingers around and under the object until I could ease it out of the ground. I lifted it to my nose but even before the scent filled my nostrils, I knew from the way it was attached to the roots of the tree that I'd found a truffle – about the size of a squash ball, with a coarse, ochre yellow skin like an unpeeled lychee. It had an intense, garlicky scent, and when I cut it in half, it was hollow in the middle. There were more there too, like a cluster of underground grapes.

I was really excited. It might not have been the prized English summer truffle, but it was still a truffle – a real one – and I'd found it all on my own, so I was ridiculously pleased with myself. I hurried home, got out my precious book and began the process of identifying what I'd found. It wasn't easy, since all the illustrations were in black and white, but in the end I was confident that I had dug up a Genea truffle. It was the right size, shape and colour, and it had the right scent, so it couldn't be anything else, but until then, no one knew you could find them in Yorkshire.

It turned out that the whole wood was full of them. When I got home and told Mum, she insisted we go straight back out and start truffle-hunting together. We found them practically everywhere we looked – we had happened to hit just the right time of year for them. We searched under promising trees – birches, beeches, oaks and hazels – prodding with our dibbers like squirrels burying nuts, about six feet from the trunk and about three to five inches down, and we struck gold more times than not.

Since that day, I've learned that it has to be that same time of the year if I'm going to find truffles there, but until recently I kept that information a secret. This year, I did finally tell one chef about the Genea truffles and when to find them in my local wood but, since they weren't a commercial variety, I didn't think I was letting him in on a fortune. Over the next two weeks he went into the woods, following the directions I gave him, and finally one Sunday morning I got a text message: 'I didn't find any of the yellow truffles but I think this might be a truffle?' With it was a picture of a very ripe summer truffle or a new autumn truffle. Whichever it was, with its jet-black colour and white marbling, it was definitely a valuable truffle. I was ecstatic and irritated at the same time: I'd always suspected that there were valuable truffles in those woods but I wanted to be the one who proved it!

A few days after we'd found the Genea truffles, Mum and I drove down to Wiltshire and rented a little log cabin in the Savernake Forest. I'd read an article about a boy who used to go into the woods barefoot and feel for truffles under the soil with his feet, and I thought, 'I've got to try this.' His parents owned a Michelin-starred restaurant, the Little Harrow at Bedwyn, and if the customers were interested, he'd take them truffle-hunting after lunch. I couldn't afford to eat in Michelin-starred restaurants, so Mum and I hung around outside, hoping to tag along.

Sadly, nobody wanted to do it that day, so we had to go searching on our own. We never found a single one – but we did find a huge patch of delicious ceps. They are traditionally called penny buns in Britain because that's what they look like, but they're also known by two different names: ceps if you're French and porcini if you're Italian. I tend to call them ceps because then my chefs know what I am talking about.

It was glorious, an area the size of Mum's garden, which was quite large, absolutely full of magnificent ceps. I said to Mum, 'You know what? I could sell these!'

She gave me a sideways look. 'Really? Who to?'

It was a very good question and I had to admit that I had no idea, but my mind was racing. I knew now this was going to be my calling – I'd finally found my niche in life.

I took a couple of ceps for us to eat that evening – they have a fabulous, slightly smoky, chestnut flavour – but it was pointless to pick any more. They would have withered and died before I could have found a buyer for them and I was well aware that if you want a sustainable business – and a sustainable world for that matter – you always have to leave enough of whatever you are picking to allow it to regenerate. So I thought to myself, 'I know where they are and if I don't touch them, they'll grow again next year. So I need to go away now, work out how to sell them and then I can come back.'

Despite not finding a single specimen while we were away, my truffle obsession had now become so strong that when we got back from Wiltshire I decided the next step was to grow some of my own. I'd heard of an expert, Dr Paul Thomas, who was doing some very ingenious stuff. He put beech and oak saplings under a microscope to make sure there were no other mycelia – the umbilical cords for mushrooms – on their roots and then, putting it in layman's terms, he stuck them in a bucket of manky old truffles, to inoculate the roots with spores. I bought six online for £150, so he was definitely on to a winner, though I wasn't sure I was at that point.

So I'd got my saplings, and all I had to do now was find the perfect place to plant them and then sit back and wait for the half dozen years it would take for them to grow large enough to produce a harvest. I couldn't plant my trees on my own land, because I didn't own any and Mum's and Grandad's gardens were already full of trees, so I had to find a suitable place that was either on common land or had a sympathetic owner who'd give me permission to plant them. So I headed out into the fields with my soil-testing kit, knowing that I had to find a south-facing slope on a limestone hill with the right soil pH, relatively sheltered and with plenty of sunlight. I decided that just above the clanging gate would be the best place, taking care not to step on any wood wasps' nests. I knew the owner, who lived in Scotland, so I took a chance, planted them and hoped for the best.

I'd made sure there were no other trees nearby, because otherwise the mycelia on their roots might have overpowered my truffles before they'd had a chance to develop. It could easily happen.

In 2007 Asda announced that they would be planting some truffle saplings around their store in Pontefract in Yorkshire, in a trial to produce affordable truffles for their customers. Before they started, they did do a bit of research but probably not thoroughly enough, and maybe they didn't look after them properly. I'm sure they also didn't realise that Pontefract is covered with liquorice root – that's why the factory that made Pontefract cakes (though we always called them 'Pomfret cakes') was sited there. Liquorice root permeates the soil and prevents other plants getting established, which could well have stopped Asda's truffles growing. It would be great to get homegrown truffles in Asda, but thirteen years later I still haven't seen one.

Bearing those kind of mishaps in mind, I had chosen the spot for my truffles with great care. I planted my saplings and then went back to check on them every few months. They looked a bit stunted at first, because once you've inoculated the roots they become more interested in producing truffles than in growing the tree, but before long they were taking off.

I nurtured those trees for years. After I had moved in with Chris in Doncaster, Mum took charge. By August of the sixth year I was keen for Mum to get her skates on, and her truffle dibber at the ready, because I reckoned we should have a good chance of finding some. Back in Doncaster, I was waiting for her to get back from truffle hunting and call me from her home phone; at that time Mum never rang anyone from her mobile with it still being a 25-pence-a-minute job. So when I saw her mobile number come up I was really excited. I answered straight away and Mum told me that after all these years of nurturing . . . no truffles.

When she'd got up to the field she'd discovered that some horses had been kept in it and not only had they trampled the young trees but then rabbits had gnawed their bark and killed them. Well, I'd given it a good go. I did keep checking the ground around where the trees had grown for a while afterwards, but it's been a few years since I last looked. Maybe later this year I'll get my dibber out and have a poke about. Mycelium is a strange and wonderful thing, and maybe, just maybe, I'll get a truffle crop there one day. But I'm not spending any more money on truffle trees.

NOBODY IN THEIR right mind would consider picking weeds for a living as an occupation. And neither did I. For me, becoming a professional forager happened organically. It took a bit of luck, a lot of knowledge and a huge amount of belief in myself. A boy I knew from school had become a very successful worm farmer and I thought, If he can make money from worms then I must stand a fair chance from weeds.

Inspired by my discovery of the ceps, I started searching for other kinds of mushroom and soon found what was to become my calling card. I discovered that there was a disused coalmine near our house – there are lots around the area, as Doncaster is on the edge of the South Yorkshire coalfield. When the pit closed, the spoil heap had been levelled and then covered with topsoil. Trees had been planted, wild birch and a whole variety of flora and fauna had sprung up to colonise the site, and it was now covered in giant puffballs. The fresh, uncompacted matter must have been full of spores, and was ideal for spreading their mycelia.

I'd first encountered a giant puffball in my early teens. My friend Paul and I were nicking off school and we decided to keep out of sight and walk along the disused railway line between my house and my Nana and Grandad's. We came to a bridge and beneath it was a single, giant puffball mushroom. It was a belter, the size of a beachball. Neither of us had ever seen one before, but there it sat, pure white among the remnants of black coal at the side of the track. I remember we were both surprised at how white it was. Although I hadn't come across one before, even with Grandad, I instinctively knew it was a fungus. We should've taken it home with us, but instead, we did what all kids tend to do with giant puffballs – we gave it an almighty boot. To our delight, bits went flying everywhere as it disintegrated into a thousand pieces. The memory has always stuck with me – little did I know then that that mushroom would one day be a very significant thing to me; that giant puffballs would change the course of my life.

I'd found the dregs of the previous year's crop a few months before. An old puffball is a miserable, sorry-looking thing, like an old wasps' nest that has been kicked around the woods. It's not a good idea to disturb them, because they aren't called giant puffballs for nothing. They give themselves every chance of survival by growing as big as they possibly can to produce the maximum number of spores. They contain about three billion, on average, and they're all looking for a new home, which preferably should not be the lining of your lungs because they are highly carcinogenic.

That's why you should never step on a common earthball either, as they are said to be responsible for more poisonings every year than any other mushroom. The common earthball is a brownish yellow or orange, with a hole that spurts out black spores that are itching to find a nice moist home, but again preferably not in your lungs. Every year a handful of chefs will get in touch to show me a picture of a common earthball, all excited because they think they've found a truffle as the interior of the mushroom is black and compact. Of course I'm quick to put them on the right path.

Before reaching such a sorry state, puffballs are absolutely delicious, and this year's crop was now ripe. If it sounds like an inflated basketball when you tap it, it's ready to pick. I gave the nearest a slap with the flat of my hand, and it pinged beautifully. They were absolutely enormous. An average puffball is the size of a small melon but the really big ones – like these – are enormous. The only thing I could find that was big enough to put these in was a box I'd had a washing machine delivered in! Giant puffballs always grow better where the local council has given the grass a good mowing, which was clearly the case here, but they are true wild mushrooms; no one has ever been able to cultivate or propagate them . . . or not yet, anyway. When I got home that evening,

I looked up the world record size on the internet and I was thinking, 'I could beat that morning, noon and night with the ones I've found round here.'

I have had a hankering to try to cultivate puffballs for a while and I reckon I could probably do it. I've got the on-the-ground knowledge, I've learned enough science and I've got the kit: petri dishes, microscope and a HEPA (High Efficiency Particulate Air) filter to take spores out of the air. About three years ago I even went as far as emptying the under-stairs space at home to make a lab. I painted it white and covered everything in plastic so it was super-clean. Then I discovered we had mice. I'd bought sawdust as mycelium food but before I started using it, the mice had found it and set up home in it – not a good start. So now the under-stairs cupboard is back to being somewhere to shove our junk. Maybe one day I'll have time to try again . . .

I'd taken some pictures of the puffballs I'd found, so I posted them on a nerdy mushroom website I had joined. A guy immediately sent me a message through the site, just saying: 'Call me.'

I knew nothing about him, but after a day or so I thought 'why not?' and got in touch. The guy turned out to be a professional forager, who I've called Mushroom Martin ever since. He was in his fifties, very tall and thin, with blond hair, a weather-beaten face and a strong Norfolk accent. Mushroom Martin proved to be a very good person to know. Given how knowledgeable he was about the practicalities of foraging, over time he became something of a mentor to me.

He was supplying restaurants and wholesalers with the giant puffballs that he'd found near his home in Norfolk and had an incredible reputation, but he couldn't meet the demand.

'How many can you get?' he asked when I phoned him.

'I don't know exactly,' I said, 'but quite a lot.'

In the end I supplied him with six hundred kilograms, and that was only a fraction of what was there, because for every one I took, I left ten behind to produce the spores for next year's crop. When you pick puffballs you have to be careful not to damage the mycelia, which are as thick as pencils, with a network that can spread for miles underground.

Before we said goodbye, Martin said, 'And do you have any ransoms – wild garlic – near you?'

'Yes, any amount. We're knee-deep in it.'

'Well, when I come to get the puffballs, can you get me two hundred and fifty kilos of wild garlic as well?'

The Latin name of wild garlic is *Allium ursinum* – *ursa* is the Latin word for bear. It is so called because in Eastern Europe the brown bears are so fond of the bulbs that they dig up the ground to get at them, and wild boar root them up too. I knew a local wood that was full of it, and when I went to see the owner and asked for his permission to pick some, he just said, 'Oh, thank God, I'm sick of mowing it, so help yourself, the more the better.'

Chris and I went straight out. It was a steep learning curve, but we learned a great deal from Martin. He explained that preparing wild garlic to be restaurant-ready is completely different from grabbing some for your evening meal. It has to be picked by hand, laid out neatly and packed in a certain way. You can't do it mechanically, partly because the places it grows tend to be in woods, but also because your harvest has to be completely free of any other plants, especially dog mercury, ivy, lords and ladies, and Jack-in-the-pulpit. Jack-in-the-pulpit is poisonous but its leaves are very similar to wild garlic, as are the leaves of lords and ladies, so you have to be particularly careful – it's quite important not to give restaurants something that's going to kill their customers! For obvious reasons, Martin was very particular about us supplying him with wild garlic leaves and not with lords and ladies or anything else, although despite our best efforts, the odd wood frog did get in, but they were usually alive and kicking so Martin let them go – it's heavy on wood frogs in these parts.

Martin duly took delivery of the puffballs and the wild garlic, pronounced himself happy, and then said, 'Can you get me another half-ton of garlic for Wednesday?'

We could and we did. It was the first stage in building our foraging credentials.

CHAPTER SIX

PUTTING DOWN ROOTS

THERE'S NO TRADITIONAL PATH to becoming a professional forager, and mine has been one of learning on the job while using the wealth of family knowledge I had already. But my experience with Mushroom Martin had shown me there was a real demand, and what had been an interest and a sideline since I'd returned to Yorkshire started to take shape as a serious business proposition. I was on a massive mission to track down plants, mushrooms, berries and nuts in decent enough quantities to make it viable. At first Chris hadn't taken much notice of what I was doing, but when I started getting orders for half a tonne a week, he could see the potential too. So when he was offered redundancy he joined me and Yorkshire Foragers was born.

This meant I really had to ramp it up. There were two of us to support and only nine months of the year in which to make a living. We usually didn't have to travel far to gather the edible wild plants we needed, partly because the area around our home hasn't really changed since the demise of heavy industry, coalmines and engineering works. The extensive areas of woodland that surround Doncaster have remained largely untouched since ancient times and we still have an eight-hundred-acre wood behind our house. There are a host of virtually untouched heathlands and peat marshes around the Doncaster area, with flora and fauna that have adapted to and flourished in those very specific habitats. For the most part, the big landowners haven't developed their huge holdings, so the hedgerows are mature and richly varied – the more plant species, the older the hedgerow. There are even exotic trees that you wouldn't expect to find in Yorkshire in a million years. For some unknown reason the Town Council decided to plant lots of fruit and nut trees during the 1950s, so among others, we've now got mature almond and Morello cherry trees growing wild all over the place.

At the start, Martin took a big part of my business but, mindful of not putting all my eggs in one basket, I gradually developed my own contacts in the restaurant trade as Martin's turf was Norfolk and certain parts of London, where there are hundreds of restaurants and Michelin stars galore. There is also New Covent Garden Market where he could go any day of the week and sell wild foods in bulk to wholesalers. In the north, there were a lot fewer high-end venues, and, through sheer naivety, my attempts to get a toehold in the wholesale market was almost a monumental cock-up.

When I went to New Smithfield Market in Manchester to see all the main wholesalers about foraging for them, they were really positive and I went home happy, having made verbal agreements to supply each with wild garlic and other foraged plants the following spring. However, when the time came and I got in touch to arrange delivery of the first batches, I was given the cold shoulder by them all.

'I'm not buying from you,' one said to me, 'because you've gone round and talked to the other wholesalers as well.'

I now discovered – too late – that the wholesalers' etiquette required me to deal with one outfit exclusively. It was a bit like an antique dealers' ring. The wholesaler would eliminate any risk of his competitors driving prices down by setting the level at which he'd sell to his retail customers. Then he would agree to supply his competition at a discount that enabled those wholesalers to sell to their customers at the same price too. Result: happiness . . . for the wholesalers at least. Having unwittingly breached the unwritten rules, I was now an outcast. Buy from me? They wouldn't even talk to me. I was persona non grata. I was distraught; I'd had visions of tootling in with hundreds of kilos of wild garlic and now it wasn't even going to be one kilo.

Fortunately Chris Herstell from Bettaveg, one of the wholesalers, stepped in to help me out and smoothed things over. I'm forever grateful to him and have been a regular at Smithfield ever since. Another wholesaler, Organic North, did some research with the Soil Association and discovered that they could buy my foraged food as organic, which also helped my business. It did take everyone a while to get over me being a woman though, but now they respect me because they know that I'm a professional. These days there's invariably a bit of a welcome party when I drive in; people emerge for a coffee and a chat – which I'm always up for – and I make sure to pop in on Chris.

WHEN I FIRST started out, I also wanted to build up a list of restaurants to sell to directly. But I knew that if I set out knocking on doors without a track record, it would be difficult to persuade chefs who didn't know me from a bar of soap that I had the knowledge of wild plants and the foraging skill, not only to get them unusual wild ingredients but also to guarantee their continuity of supply. This time it was Mushroom Martin who gave me a helping hand to get started.

'Tell you what,' he said, 'I've got a friend, Aiden Byrne, who has a restaurant in Manchester. He was twenty-two when he got his first Michelin star, the youngest chef ever. Go and see him and tell him I sent you.'

So I made an appointment and arrived dressed to impress, in my usual full hair and make-up and wearing an orange spangly top. Aiden's expression when he saw me suggested that this was going to be a very hard sell and he gave me the full third degree, testing my knowledge of wild foods and how I would source them to ensure the quantity and quality he wanted. It went on for about an hour, but at the end of it, he sat back and smiled.

'I've got to be honest, Alysia,' he said, 'if it hadn't been for Martin's recommendation, you wouldn't even have got through the door. I can't see the forager, but now I've heard the forager, let's give it a try. Can you get me some wild garlic capers?'

'Absolutely.'

'Good. I need as many as you can supply.'

'Are you sure? I can get a lot of them.'

'I don't care, just get them. I can use them, but you've only got three days to do it.'

We agreed a price per kilo and I went off to gather them. Now wild garlic capers are amazing and expensive little things and I mean *little*. After the flowers of wild garlic have turned they will eventually become a seed pod. There's a short window, about two weeks after the flower dies, when they are a tiny green immature seed pod minus the seed, and this is when you need to pick them. They are about eighteen inches off the ground. It's back-breaking work because of the bending, although it's a bit better if you can start at the bottom of a slope and work your way up. They weigh next to nothing but have an intensely garlicky flavour. I turned up on the specified day with twenty kilos and since a single caper weighs less than a peppercorn, that meant thousands and thousands of them. When Aiden saw them, he just said 'Bloody hell!' but was as good as his word, paid on the spot and set his chefs to preserving them. Aiden became my first Michelin-starred customer and all these years later, I'm still his source for foraged ingredients.

Having landed the contract with Aiden, I sat down and thought, 'What's the natural progression from here? Which other chefs can I try?' The obvious next step seemed to be to talk to Simon Rogan, then at the French restaurant at the Midland Hotel in Manchester, as well as the restaurant that had made his name and gained him two Michelin stars: L'Enclume at Cartmel in Cumbria. L'Enclume was famous for its use of foraged ingredients, so I felt I would be pushing a door that was at least half-open. I arrived at the meeting once more wearing my trademark sequins and probably too much make-up and Simon's head chef, Adam Reid, and a couple of his other chefs gave me a real grilling to see how good my knowledge was, as obviously they wanted to be sure that I really knew what I was talking about, before trusting me to supply them. Fortunately I passed the exam.

More recently, the Simon Rogan connection also put me in contact with Mark Birchall. After a spell of over five years as head chef at L'Enclume, in 2017 he launched his own restaurant at Moor Hall in Lancashire. Since then he has earned two Michelin stars and in 2019 Moor Hall was crowned the

the Estrella Dam Restaurant of the Year. I have been providing him with foraged ingredients since the restaurant opened. When my chefs get these huge accolades, I am always proud to have played a minuscule role in their success.

Closer to home, I'd heard of Winteringham Fields in north Lincolnshire – an acclaimed restaurant owned by Colin McGurran who had won twice on the *Great British Menu*. So one day I called the restaurant out of the blue and got through to their head chef Slav. It turned out Slav – short for Slawomir Mikolajczyk – was from Poland and so my heritage went down very well. We spent about two hours talking on the phone about foraging and everything Polish. Having hit it off straight away we have since become good friends. The day we first spoke he was very excited as he had spent the entire week working on textures of onions. As it turned out he was already trying to find a forager and had even contacted his local council to see if they knew of one. So selling the services of Yorkshire Foragers to him was a much less daunting process.

At Winteringham Fields it really is a Winteringham family – it's just one of the greatest places to work, and the team all have such loyalty to Colin it's easy to see why. I used to turn up after coastal foraging with sea herbs fresh from the beach, not even washed. Slav would go through the bags and pick out what he wanted. He was always keen to know what I could find and what was coming into season – his passion for food was like no other.

I went to his wedding last year and got talking to some of his relatives who had a similar story to my grandad. It took them longer to get out of Poland because the invading

Russians had sent many of them to the harsh gulags of Siberia. So Slav and I have a lot in common.

Slav is a truly amazing chef and now has his own place, the Hope & Anchor, a mile down the road. He arrived in the UK without a word of English and owning just a bicycle, but he is now a successful business owner and community employer with a huge reputation. A great Polish immigrant success story. The Hope & Anchor is my Friday afternoon haunt whenever I can find the time, because there is nothing like looking out on to the Humber estuary with a glass of wine in my hand, and knowing that I have the next couple of days off. The food is excellent too and I always have the slightly sneaky intention of treating myself without telling Chris.

With an established customer base of restaurants and wholesalers, I had the idea of conjuring up some 'Christmas Tree Syrup' from Sitka spruce needles. Sitka tips are the new growth on Christmas trees, they are bright green and full of sugar and softnesss. Mum had taught me to make jams and syrups from foraged ingredients when I was young. The syrup had a lovely citrusy taste and I thought the seasonal connection would make it a no-brainer for retailers to stock it by the barrel-load. So I made up a batch and took it up to my friend Craig Atchinson, who I'd met supplying the Grand Hotel in York, to get his opinion. Craig really loved it, in fact he was so enthusiastic he made some designer cocktails from it right away, which went down very well. I knew Craig would have told me truthfully if it was awful, but it wasn't. His verdict felt like a good sign, and the first hurdle out of the way. I then got a designer friend to produce a nice label and sent samples on spec to the kitchenware company Lakeland who sell a range of high-end foods in their store in the Lake District and around the country, and online.

To my surprise and delight, the buying team at Lakeland loved the syrup and were ready to place an order at once, but only after I'd proved to them that I had complied with all the standard food safety and hygiene requirements that large food retailers are required to follow. They hit me with a sixty-four-page document containing what seemed like a thousand different rules, regulations and specifications. The regulations covered everything that had been used, from the clothes worn when gathering, preparing and cooking the ingredients, right through to how the containers had been sterilised and the temperature points at which the product had been stored, prepared, cooked and cooled. Having read through the document – a time when my legal training really came in useful – I realised I was unable to tick a lot of the required boxes because they simply couldn't be applied to wild ingredients.

I contacted Lakeland to explain my problem and very generously they offered to fly their food technologist over from France to advise me and steer me through the process. Karen turned out to be a very nice Englishwoman in her mid-fifties, with dark hair piled in a bun. She lived and worked in France, but was a consultant to several British companies and had an encyclopedic knowledge of food hygiene rules and regulations.

Having painstakingly worked our way through the paperwork, we came to the government regulations produced by the Food Standards Agency. It was mind-bogglingly technical, but only four pages long, so I thought, 'If I can do a law and politics degree, surely I can do this?'

Somehow I got through it, but next came the specifics of my particular product: the Christmas Tree Syrup. The pine needles were absolutely natural and untreated, and Karen was concerned by the number of pathogens they might contain. I pointed out that with the exception of pine martens – which in any case are very rare in England – most wildlife avoids pine trees. Admittedly squirrels do eat the pine nuts and this gives the tree its best chance to propagate itself – they are untidy eaters, and drop a few fertile nuts on the ground, or bury them for winter and don't find them again. Everything else is deterred by the needles' tough outer casing. Even bugs stay away, because they contain a natural insect repellent – I can be picking them all day long and I never get bitten. And birds are not keen to land on pine trees, because the needles are like the spikes that keep pigeons off the windowsills of office blocks. But Karen needed something more scientific than that.

To get rid of any pathogens, she wanted me to do a hot infusion of the needles – boiling them up. But when I tried to do that, the needles disintegrated and, even at just thirty degrees, the liquid became a murky swamp. The only way to achieve the clear liquid I needed was to do a cold infusion. At a maximum of eight degrees, the liquid was crystal clear and, by the way, really delicious, but to be sure that no pathogens had been introduced into the sugar mix, which could have caused it to ferment, I then had to add a preservative – citric acid – and potassium sorbate, a mould inhibitor used in wine-making, which I discovered the hard way *must* be used within its sell-by date.

Having done all the hard work to make the syrup tasty and safe to eat, I still had to prove it by sending samples to a lab for analysis. Commercial labs charge a fortune for this, but luckily the University of Chester runs an excellent, well-respected and, best of all, very inexpensive food testing lab as part of their food technology courses. For a mere £4 or so per single test, they do a range of tests that tell you whether the foodstuff you're producing is safe to sell. So it's cheap as chips and it does actually give you peace of mind, and if you don't test the ingredients that you've foraged, you're running a huge risk.

Amazingly enough, back in the day when there weren't any supermarkets and we didn't have sixty-four-page documents to fill in, people just went ahead and made those kind of things, and people bought and ate them and hoped for the best. But today, if a retailer is going to sell any foraged wild food to the general public, they want to make sure that there is zero chance of any health scares or potential public relations disasters as a result, so they want to be certain that every single box has been ticked and double-checked before they'll stock it. The tests came back all clear, so at least I knew nobody was going to keel over from consuming the syrups.

I still hadn't finished satisfying the myriad requirements. I still had to sort out barcodes and VAT codes and even find the right recycling label. And just try getting insurance on foraged food. I had to get a specialist underwriter and even then it was all a bit vague, but good enough to tick the insurance box.

Even buying caster sugar turned out to be less simple than you'd think. What I didn't realise is that drug dealers buy it in bulk – since one white powder looks much like another, they presumably 'cut' cocaine with it – so if you try to buy a large quantity, you may get a visit from the local drug squad. Luckily my contract with Lakeland provided the evidence that I was buying sugar for a cocktail syrup and not setting up a narcotics empire, and the drug squad's hopes of arresting Doncaster's answer to Pablo Escobar went unfulfilled. It still meant I was on the local police records as a supplier of caster sugar, but that doesn't have quite the same ring to it.

Eventually I'd ticked all the boxes. As December approached, the Yorkshire Foragers Christmas Tree Cocktail Syrup duly went on sale and did very well for us.

It felt like another obstacle overcome. No matter how tough it had been – and it had been really tough at times – I had enjoyed the process of supplying the syrup and we'd made a profit. I'd been through a tremendous learning curve. and had gained invaluable experience. I was proud to have singlehandedly gone through the entire process from start to finish and to have made it work by thinking through my own solutions and through communicating with people. To run a successful business you have to make it work, you can't just stop, you have to follow through with what you've said you can do. I didn't want to get back into supplying retail again straight away, but it's something that I can return to in the future and maybe even try on a bigger scale.

IN THE MEANTIME, supplying wholesalers and restaurants is the mainstay of Yorkshire Foragers. These days I can quite happily ring or call on any chef and when they ask who else I supply I am able to reel off a string of well-known names.

Name dropping goes down well in the restaurant industry, but not for the reasons you might think. It's not a case of being a billy-big-pants, it's because they need to know your credentials. So if I tell them I've been supplying a top chef for some time, I know that as soon as I have gone they will be on the phone checking me out. It's a very small world this industry and you only have your reputation to go by.

Sometimes that reputation means that chefs seek me out, which is a huge honour. Last year a very, very famous chef phoned me and humbly asked if it would be possible for Yorkshire Foragers to start supplying him. I couldn't believe who I was talking to and was a bit starstruck. Of course the answer was yes. I still haven't plucked up the courage to ask him where he got my number from though, or who recommended me. Maybe one day when I feel a bit braver. . .

He wasn't the first world-famous chef I've supplied though. In spring 2018 I'd been picking some pink European larch cones. They eventually become mature cones, but when they're new they have a lovely citrus flavour and are absolutely beautiful to look at. I'd posted some pictures of them on Twitter and René Redzepi, the head chef of Noma in Copenhagen, one of the world's finest restaurants, contacted me at once. That was something of a major surprise, but he'd been following me on Twitter – very flattering considering he had once been voted the world's best chef and Noma had been voted the world's best restaurant four times.

René was born in Copenhagen to a Danish mother and Albanian-Macedonian father. They spent a number of years of his childhood living in Macedonia with his father's extended family. It was in Macedonia that he learned about food and foraging and today Noma is famous for its use of wild ingredients. The Noma team gather a lot of their ingredients from the Danish forests and coasts, but René needed to know if I could get him some pink larch cones. The European larch is native to the mountains of central Europe – the Alps and Carpathians – and northern Poland and southern Lithuania. They don't grow in Denmark, but a lot were planted in Britain from the seventeenth century onwards, so they're relatively easy to find here if you know where to look.

Supplying one of the world's best restaurants was an amazing thing for me, and I offered to send some over straight away. I then got in touch with Aaron Goldstraw, the head buyer at Oliver Kay, a £60-million pound fruit and veg company with a fridge in their warehouse the size of a football pitch.

Aaron had the toughest start in life imaginable. He told me about the day he was taken to a children's home. He and his mum had been out knocking on neighbours' doors asking for food and came home to find social services there to take Aaron and his brother and sister away. He was just six years old, his brother was three and his sister not even yet one. Luckily all three children were then adopted by the same foster parents, who brought them up together in a loving and caring home.

Aaron has achieved great things, but his generosity and sense of giving back to those who are less fortunate is enormous. He is often to be found at charity food events just doing the washing up, and he is a proud member of the Southport Food Bank, giving up part of his weekends to cook for the homeless. Only the best will do, so it has to be one of the best soup kitchens in the country. He picks the best produce the warehouse fridge has to offer, using all the luxury ingredients the rest of us mere mortals can only dream about. The meat comes from Booths, known as the 'Waitrose of the North'; there are Roscoff onions, and Southport's homeless don't eat flavourless, hot-housed Dutch tomatoes, but hand-picked ones from the volcanic slopes of Mount Vesuvius and the valleys of San Marzano – a classified fruit of Italy.

Aaron was generous to me too, giving me a chance in the early days of Yorkshire Foragers to supply Oliver Kay, even though I was an unknown newcomer to the trade, and without him I wouldn't be where I am today. I became the wholesaler's forager and still supply them now. Aaron and I have a very good working relationship and are good friends as well. He is a real foodie, and since Chris doesn't see what the fuss is about when it comes to fine dining, if I get invited anywhere posh I often end up taking Aaron, who, like me, is always up for a fabulous meal, especially if there's a wine flight involved.

Oliver Kay were trying to raise their profile, and had asked for my help. I'd already introduced Aaron to Mark Birchall over at Moor Hall, and when René got in touch, I said, 'If you really want to tap into the top end of the market, René Redzepi at Noma in Copenhagen wants some of my pink larch cones, so perhaps we should do it together. The publicity will be good for your business and for mine too. Why don't we fly some over to one of your agents in Holland and then they can drop them off up there?' Aaron really wanted to go and meet René, so he decided to fly over himself and take the cones in his hand luggage.

I messaged René and asked him if it was all right to tweet the hell out of it; he was happy and so was I. So I organised the whole thing and then we bombarded our Twitter followers with pictures of me picking the pink larch cones and handing them over to Aaron at Oliver Kay's HQ in Bolton, and then got yet more snaps of Aaron turning up at Noma and handing them over to René.

Aaron told me later, 'When I walked through the door, René couldn't believe I'd gone all that way to deliver them in person.'

'And did you get a fabulous dinner out of it?'

'No, damn it! It's a six-month wait for a booking there and he couldn't squeeze me in, though he did show me round the restaurant before it opened, so it wasn't a completely wasted trip!'

By the time Aaron had finished looking around Noma, his taxi driver had got fed up with waiting, so Aaron had to walk the five miles back to the city centre. He then got caught in a thunderstorm, but it didn't dampen his enthusiasm and he was still buzzing about the trip when he got back. While he was in Copenhagen, I found out that another friend of mine, Nigel Haworth, a Michelin-starred chef at Northcote, happened to be there as well, so I put the two of them in touch and they spent the next two days eating and drinking their way around Copenhagen together.

René was so pleased with the pink larch cones that he posted me a message that now heads my Twitter account: 'Thank you for supplying us, without people like you, Noma simply wouldn't exist.'

It doesn't get better than an endorsement from René

Redzepi. With the help of endorsements like his, I have built a reputation for supplying top quality ingredients and for absolute reliability; if I promise a delivery, I make sure the chef always gets it on time. In a small world where everyone knows everyone, you're never going to get away with missing delivery dates, nor with ducking and diving, scamming your customers or delivering inferior produce – not that I ever would. And at Yorkshire Foragers, we wouldn't have it any other way.

CHAPTER SEVEN

CHANGING SEASONS

TEN YEARS SINCE we officially launched Yorkshire Foragers it is thriving in ways I could never have imagined – who'd have thought I'd be on the telly? I still won't watch myself on it, I prefer to think I'm just having a chat with one of my chefs in their kitchen rather than talking with thirty Wembley stadiums full of people watching me. But those days are rare, and most of the time you'll find me – and the rest of the family – tramping through woods and across marshes searching for weeds. That definitely takes away any delusions of grandeur and glamour.

I told Chris from the beginning that Yorkshire Foragers is a dictatorship not a democracy, and these days he's happy to go out with my instructions and forage what we need. But it took him a while to get the hang of it. When he first began working with me, I said to him one morning, 'Can you get me some sweet chestnuts?' and he came home with a ten-kilo bag of conkers. Ground-up conkers were once used as a medication for horses with breathing problems, but they're toxic to humans and definitely not on the list of delicacies we supply. They aren't that good for dogs and I can't imagine they were that good for the horses either.

Another time, he came home one lunchtime, very proud-faced, with a sack full of wild horseradish roots he'd spent an entire morning digging up. Having taken a quick look, I had to gently point out that they were apple tree roots. On long walks with our chocolate labrador, Fred, I gradually taught Chris everything that Grandad had taught me, and everything I'd learned for myself, and before long he was up to speed and playing a full part in the business, including being our champion puffball spotter.

Chris and I are both out most of the year, although it's a bit harder to get him out in winter when he would much prefer to stay indoors. In winter he has the central heating set so high that our house is like Barbados. I wish it really was Barbados but there isn't that much money in picking weeds – if we are lucky we can make it to Lanzarote and pretend we are warm. As soon as I see any sign of spring growth, I turn the heating off when Chris isn't looking so it's more like Bournemouth than Barbados.

Chris is not a great one for customer interaction, not always being top of the class at tact and diplomacy. That was illustrated when he told a young lady who worked in catering that she ought to try hospitality instead, so he tends to keep firmly on the foraging side of things these days, while I do the front of house stuff. He's not a fan of fine dining or fancy food, and unlike me who likes to be tactful, he'll feel the need to be honest when he tries something he doesn't like. Now, I've had to try some real shockers in my time, but whatever it has been I've never let on I didn't like it. I've always left saying, 'Thank you very much. Yes, it was lovely. Thanks.' The chefs are my customers and I wouldn't want to offend them, even if they do make me try deep-fried calf brain in miso crumbs.

My mum, Barbara, is a fantastic forager and can still hack

a day on the wooded slopes picking wild garlic, or up on the moors with her special bilberry comb, or tootling down towards Sowerby Bridge looking for sweet cicely. She will even help out on the salt marshes at the beginning of the year, when it's quite tough-going. When you're out and about you get an eye in for certain plants, and Mum is like that with lingonberries; whenever she finds a new patch, she gets all giddy kipper about it and has to phone me straight away so she can describe it to me at great length. I have a mental map of everything we do, so when I ask her where she's found them, the conversation will go something like:

'Well, you know where the number two hump is? When you get past the canyon?'

'Yesss . . .'

'Well, you go on the ridge path up to the holly tree with no berries on and down that path where that crazy dog woman goes with her six labradors with kennel cough.'

'Right . . .'

'Then it's just past that big puddle and on your right by that birch tree you took the bark off the other year.'

And I'll know exactly where she means.

I think I keep all my family busy looking for things, whether it's Mum with her lingonberries or my brother Adrian, who is really good at picking wild garlic now. Adrian works for a brewery, delivering beer and he always keeps an eye out for mushrooms while he's out and about.

Adrian's son Ryan particularly enjoys going puffballing, and I like to keep him on the lookout for a good supply of acorns. I also take him to help me trim the tops of heather up on the moors and get his older brother Alex picking Sitka tips in spring. Ryan can't give my mum a run for her money yet – she's four times his age but five times faster – but give him a few more years and we'll see. I think when my nephews are older, they will come to realise the significance of foraging and how brilliant a job it can be.

We've had a few truffles recently and Ryan has surprised me because – despite being a picky eater – he really likes them and has become a bit passionate about them, like me. But we both have a guilty secret: we like nothing better than to open a can of Tesco's meatballs in tomato sauce at 87p a tin and then shave about thirty quid's worth of truffle on to them and sit in front of the telly while we eat them. I know . . . we should be shot for the sacrilege. But when you've done the scrambled egg and truffle thing, then the truffle in honey for the cheeseboard, and had it on fresh pasta, you start to run out of truffle ideas unless you want to make something really fancy, and I save that for the chefs. So truffles on Tesco's meatballs it is.

Young Ryan also gets taken where I take him and he has to eat the food whether he likes it or not. Recently he and I had had a day out in Cleethorpes. On the way home I said, 'Mcdonald's or pub food?' but while he was still thinking about it, I drove straight past Mcdonald's and took him to Slav's pub, the Hope & Anchor. The welcome is always amazing there, and when we walked in they sat us down and brought out some beef shorthorn ribs, truffled chips and green beans in an amazing sauce.

Ryan began to quiz me over the bones, fat, etc., etc., but I said, 'Ryan, just shut up and eat it.'

He reluctantly put one of the short ribs in his mouth and then spent the next few minutes saying, 'Oh my God, oh my

God. You said this was pub food, but it's not pub food, it's much, much better than that.'

'Ryan,' I said. 'Remember this and whenever I take you somewhere, trust me, don't ask and just eat.' He agrees with me now and we reminisce over the short ribs more frequently than we should.

OUR FORAGING YEAR starts in earnest with the spring solstice. As it approaches, the weather can still be icy or even snowing, but nevertheless you'll find me walking the long path to the woods to see if the wild garlic that's been poking through since Boxing Day is big enough to pick yet. With the worst of winter behind us, the sun starts coming back, the light strengthens, the days lengthen and the soil starts to warm up. It is the signal all wild plants have been waiting for, and for us as well. My early spring walks take me past a big white farmhouse which always looks like no one lives there, along a path by a big wheat field till I am walking past the dead elders, looking for any sign of Jews' ears mushrooms starting to grow in the damp nooks and crannies. Jews' ears and wild garlic are both familiar sights that mark the progress of the seasons, but I almost always find something unexpected as well – the broken branch of a birch tree, perhaps, weeping sap on to the leaf litter; I can reach up – though not that far, being only five foot four – and taste it; it's like pure, sweet water. Even when I was a child, I would notice the first hints of new growth, like the snowdrops pushing up through the winter blanket of snow or the faint greening of grass in a sheltered, south-facing dip. Half a lifetime later, those early signs indicate the beginning of our busy season as the long, dark months begin to recede. It's the beginning of our foraging marathon, and then we go like the clappers right through to November, though we never completely stop because even in the depths of winter, when almost everything is dormant, there are still a few roots, mosses and evergreens that we can harvest.

As the days become longer than the nights, the trees burst into leaf. There is nothing to equal the vibrant palette of early spring, with leaves so pale they are almost lemon-yellow through to foliage the emerald of the deep ocean and every conceivable shade of green in between. March sees an abundance of young shoots of gorse flowers with their extraordinary coconut flavour, and I'm also very partial to young tree shoots before their chlorophyll content becomes too intense. I want to bring the flavour of the British countryside to the table, but I don't want to serve up stuff that just tastes like grass. We also see the first hawthorn flowers. They are beautiful, but to attract the flies and other insects they need for pollination they smell like death. I have a really good sense of smell, so I'm never too keen when a chef tells me he's made a hawthorn cordial – to me it smells like dead squirrel squash. I do think that a good sense of smell is essential for a successful forager along with cracking eyesight, nimble fingers and a lot of stamina. Unfortunately I've got a very dodgy knee but being slow isn't necessarily a hindrance because you notice more.

As the month progresses, the hours of daylight are rapidly increasing and plants, like humans, all respond to that. Although I do enjoy the dappled light coming through the trees, I can't wait for the canopy to open so that the birds don't crap as much on the plants below. As the spreading leaves create pools of shade beneath them, new flowers, plants and creatures emerge. There's a lot of Jack-by-the-hedge, with its mild garlicky-mustardy taste, coming through. Sometimes we spot wood frogs emerging from their winter slumber, to enjoy the warmth of the sun with the dampness of the defrosted earth.

April absolutely flies by because the list of plants, flowers, mushrooms and tree buds and shoots you can pick then is as long as your arm. When we get into May, while the St George's and Jew's ears are still going strong, the giant puffballs and morels are coming out as well. People tend to associate wild mushrooms with autumn and it's true that you do get a lot of them then, but different varieties are available from April right through to November, or even later in some places – though, like most plants, there's a very limited period when you can pick them. St George's mushrooms are only there for three weeks, and most of the others are similarly short-lived.

Picking plants in the woods as the weather warms up is like being an all-you-can-eat banquet for bugs. Even in June, I have been known to wear a thick jumper and baggy cords to try and keep them off me, but I can see the mosquitoes sitting on my sleeves, trying their luck, and I guarantee I will still get bitten by something around the knees where the material tightens enough to let them get their nasty little prosboces into a handy vein. No amount of DEET spray will suffice, I still get bitten, even though as a general rule, I don't really attract bugs. It helps if I can just keep close to Chris, as they seem to prefer him to me; Chris is catnip for mozzies and gets bitten all day, every day. We spend an absolute fortune on Jungle Formula wet wipes, sprays and wristbands. Some of the sprays we use actually smell quite nice, as such things go; I've even been complimented on my perfume a few times and asked 'What is that you're wearing?'

'It's Jungle Formula by DEET,' probably wasn't the answer they were expecting.

AS THE YEAR MOVES ON and the leaves mature, some become tough and sap-stricken but others ooze softness and flavour. Lime leaves are delicate and sweet when fully formed and blackcurrant leaves take on the flavour of the fruit. Edible flowers now come high on my list, like those of borage – purple, with a honeyed taste, to enliven salads and bring a splash of colour to your dishes.

Although we seem to be flat out from the moment the light begins to strengthen in early spring until the onset of winter, there is a quieter period in July, bizarrely enough, when all the plants are in transition, moving towards fruit and berry season. Everything has flowered, the fresh green shoots have matured, and at the same time, alliums, garlics and the like have mostly died off around our parts.

The lull lasts for about three weeks, while we're waiting for the fruits to have ripened enough for picking, so we can make the most of the fine summer weather – if there is any – and catch up on some R and R. Doncaster isn't exactly the Costa del Sol, but it's in the rain shadow of the Pennines so the annual rainfall is barely half the UK average so we can get very hot, dry spells in summer.

Near my mum's house there is a deep and narrow valley. The steep slopes are thickly wooded and the river running through it forms a cool, green tunnel beneath the canopy of the trees overhanging the banks. It is a regular hangout for herons. We watch as they perch on the rocks, waiting for an age to pick out the trout. When you're looking down from the top of the valley you can see where the river is by watching the herons fly up and down it as if it was the A1.

Thirty years ago the river was so polluted by heavy industry that almost nothing could live in it, but now it's been brought back to life. I remember when my nan used to pick me and my brother up from school we sometimes saw dead fish floating on the surface of the river as we walked along it. It upset me even then to see the effects of pollution. Of course there is still pollution today, and fellow dog walkers have told me of times when their dog's jumped into the river to cool off, only to come out with a severe reaction to the chemicals in the water. But these days, when we walk beside it, we are more likely to see the slow, spreading circles on its surface as trout rise to snatch flies from the surface and, if we're up early enough, we can often catch the electric blue flash of a kingfisher as it raids the water for minnows, or the motionless silhouette of a heron, stalking the shallows in search of frogs or fish. These interactions always make me happy because the river's health is a fragile thing, with all its creatures dependent on each other to thrive.

Closer to home we have the River Don, and on fiercely hot summer days there is nothing nicer for the three of us – me, Chris and our labrador Fred – than to walk alongside that river, revelling in the cool shade, the plants we see and the glimpses of wildlife we catch as we stroll along, and best of all, there's a riverside pub when we reach the end of the walk.

I love those relatively quiet summer weeks when the back-breaking work of picking wild garlic is over, and we can draw breath and smell the roses – or more likely the bouquet of a nice glass of wine – before it all picks up again, and we're once more foot to the floorboards, as we enter the fruit season. The fruit and the flowers we harvest at this time of year are fragrant and sweet. There is also plenty to collect, as there is always moisture in the soil and in the mulch of the woodland floor. We start with wild raspberries and wild strawberries in the woods and hedgerows and move on to wild cherries and plums. Apart from the bugs, high summer is a lovely time to pick, and we collect meadowsweet, the delicate flowers of sweet woodruff and the beautiful vibrant lemony leaves of wood sorrel.

As the fruit and berry season passes its peak, nut season is getting under way. The trees stop producing chlorophyll and start to draw nutrients back into their roots ready for the next growing season, while the leaves take on those glorious autumn colours – yellow, gold, russet, orange and red – and begin what the Americans call with accuracy, if not much poetry, 'The Fall'. I love walking through the woods, hearing my boots rustling through the drifts of fallen leaves and releasing those familiar, musty autumn scents. Largely invisible in high summer, grey squirrels are now everywhere, frenziedly scrabbling at the soil to bury nuts as winter food. They bury so many that a lot of them will remain unretrieved and will sprout saplings the following spring – grey squirrels probably plant more oak trees than even the most dedicated arboriculturist. Unfortunately they also do a lot of harm to trees, so it is true that we need fewer of them.

There is not much time for us to watch the squirrels or admire the beauty of the season because we're kept busy gathering the pick of the crop from the woods and hedgerows: walnuts, sweet chestnuts, acorns, beechnuts, hazelnuts, and even semi-wild almonds from the trees that Doncaster

Council planted all over our area in the 1950s. I pick the nuts myself, so I know exactly where they come from and what the quality is going to be.

As the days become shorter again, I love to be out early on those first frosty days of the year, with the low sun struggling to pierce the mist, and the hedgerows thick with fruits and berries dusted with frost or dew. I can hear the thin glaze of ice crunching as I tramp through the frozen puddles on the woodland paths and though my cheeks sting and my fingers can be numb with cold, the mist soon begins to thin and disappear among the tree-tops and the sun's rays touch the ground, warming me and bringing the earth back to life.

As winter approaches everything slows down, but there are still edible plants to be found, although some of them are harder than others. Alexanders seem to grow absolutely everywhere except Yorkshire. So you can find Alexanders everywhere except in my neck of the woods. It's known as roadside rhubarb because it tastes like a more savoury version of it, and if I did have them nearby, I could pick them when nothing much else is growing. As it is I have to rely on the plants I pick in the growing season to keep me going through December, January and February.

It is not just the changing seasons and the latitude that dictate when and for how long plants are ready, because the local climate and weather can have a huge influence too. In 2018, for example, we had three periods of heavy snow in Yorkshire extending well into spring and holding back the growing season, and then we seemed to go from freezing cold to burning hot sun almost overnight, so everything came out together, in a rush. Then we went straight into drought, shortening the growing season for some plants and shrivelling others almost as soon as they'd shown their faces. However, as I'd broken my hand in seven places that year and was having to forage with a pot on my hand for six weeks, I was quite grateful that the drought had slowed everything down.

I don't have a favourite season. I love them all, including the transition from one to the next, so slow and imperceptible that you can only ever say in retrospect that one season has ended and the next has begun. Although I always feel a little sadness at the sight of the swallows lining up on the telegraph wires ready to begin their long migration, knowing that one morning I'll wake to find the wires empty and summer over, within days I'll be absorbed in the beauty of the autumn season and the different delights it brings.

CHAPTER EIGHT

ON THE SHORELINE

WHILE WE ARE OUT in the fields and woods almost every day during the foraging season, we also have to find time to gather the many plants that can be found by the seashore. So, depending on the weather, from the last week in April or the beginning of May onwards, we'll be heading for the East Yorkshire and Lincolnshire coasts.

I love to be out there on clear spring days when it sometimes seems as if I'm the only living being in that whole vast sweep of land, sea and sky. I usually go on a Monday, when the weekend warriors have left the beach and gone back to their day jobs and we can at last have it pretty much to ourselves. I can stand for minutes on end, watching the watery sunlight shimmering on the shoals and mudflats, and listening to the cries of seabirds filling the air, under those huge skies that merge so imperceptibly with the sea that you can't tell where one ends and the other begins. Maybe it's the flatness of the landscape but the skies seem bigger there than anywhere else in Britain.

Or sometimes when I'm on the mudflats I'll look back at the pier and the funfair on the shore and every now and then the wind catches the laughter of children and brings it to my ears. Sometimes the wind brings the smell of fish and chips too, and after four hours picking on the beach that smell is emblazoned in the front of my mind. There is nothing I like better after a day's hard graft in the fresh sea air than sitting in the car eating vinegar-soaked fish and chips with a wooden fork.

I love coastal foraging, and in places where there are vast expanses of flat beach, mudflats and salt marshes, like you get on the Humber estuary and the Lincolnshire coast, there are a whole different range of delights. It has always surprised me that the plants that grow on the marshes need fresh water as well as salt water; if they don't get rain, they don't grow well. The pattern of tides ensures that half of the day they are underwater and half in the fresh air, ready for us to pick.

With so many hours underwater, the sea can give them a good wash, but whether that's a good thing depends on what they're being washed with. If you're gathering sea herbs, you always have to check for sewage alerts online because there's still a lot of raw sewage being pumped into the sea. Just like shellfish, coastal plants and herbs may be affected. It very rarely happens but it always pays to be vigilant. I also make sure that the sea herbs that I deliver still have a coating of mud and sand, not just to preserve them, but because I know that means the chef will have to wash them properly.

A woman wading through the water hacking away does tend to draw attention, so I always work with my back to the shore. Away from the crowded beaches, it's less of an issue because you've got to make a lot of effort to get out there. So, even if they are curious about what I'm up to, most people won't bother plodding out over those lumpy, swamp-ridden, muddy and sand-flea infested salt marshes just to say, 'What are you doing?' especially if the answer is blindingly obvious: 'Wielding a very big machete or a very sharp knife.' When families and small children are on their summer holidays, catching some rays, splashing in the shallows or taking donkey rides, they don't necessarily want a complete stranger wandering past with a machete glinting in the sunlight, so I try to keep it hidden until I'm well past the beaches.

On the odd occasion that somebody does make it out to where I am working, their first question is usually followed by, 'Really? I've lived here all my life and I never knew you could eat that,' or, 'Is that what samphire looks like? I thought it might be, but I wasn't sure.'

We gather all the edible salt-loving plants from the expanse of salt marshes that spread out from the Humber estuary. Of these, sea purslane is one of my favourites. In April you get the new purple growth, which is beautifully succulent and tastes a bit like a salty Granny Smith apple. Sea purslane actually grows all year and never dies off, but it can get a bit ropey in winter if the frost gets to it, so by that time we leave it well alone. Having said that I've found fresh, plump sea purslane on Boxing Day alongside baby sea aster and new-growth scurvy grass. The taste of scurvy grass can be formidable, quite hot and spicy. It's a two- or three-week wonder best picked in the depths of winter when it is young and the flavour is more delicate. I don't tend to bother with it as we don't really go out in winter, and when we're back to coastal foraging in April although the leaves are edible, they're so small they're not really worth scratching around in the muck for.

I don't particularly like sea rosemary because it smells so rank that it's horrible to pick, but some chefs do ask for it. On the other hand sea blite, which also starts quite early, really is a bit of a wonder herb, with a flavour somewhere between samphire and sea rosemary, but if a chef puts 'Crab with annual sea blite' on his menu, he's not going to get many takers. I really must find a new name for it, because the traditional one doesn't do it any favours; no one wants to eat something that sounds like a disease. We also find the odd bit of sea beet, the wild ancestor of beetroot, sugar beet and Swiss chard. If you go a bit further down the coast it's everywhere, and it's quite prolific in the colder months, so there are at least two cycles a year when we can pick it. It's a really good sea veg and tastes fabulous wilted in a bit of butter, and it's full of nutrients so it's really good for you too, but that's always the case with wild food: the flavour and the vitamin content are huge.

The season for sea aster begins around the last week of April or the start of May and will then easily go on to the end of October. Aster is the Greek name for a star and the plant really is the star of the salt marsh – chefs go crazy for it when the season starts. It's a lovely succulent plant, sometimes called sea asparagus, and with good reason. Sea aster provides a source of nectar for late summer butterflies. The only drawback of this is that some species also lay their eggs on it. In boom years for moths and butterflies, every plant is smothered in tiny black eggs, which drives me nuts since they're almost impossible to remove. I don't know how they survive their immersion in salt water twice a day, but somehow they do. We only ever see the sea aster's flowers in the distance because the patch that we harvest never gets the chance to develop them, but that's OK because we don't take the roots and it's a plant that if you cut it, then it comes again. There are now plenty of growers who have learned to cultivate sea aster but I can tell from fifty paces that it's not the real thing, because the leaves are see-through and not succulent. The plant itself is just a very poor version of the wild plant, but then anything grown in a poly-tunnel is not the best version of what it can be in the wild. I am an absolute snob about it because if I ever get served cultivated sea aster, I leave it on the side of my plate in protest at them not getting hold of the genuine thing.

Sea arrowgrass, sometimes called sea coriander, comes into season in late April or May. It contains the same compound as coriander leaf, and tastes exactly like it, though you have to be careful when picking it, because it looks virtually identical to normal marsh reeds but, whereas reed stems are round in cross-section, with sea arrowgrass the clue's in the name: if you cut through it you can see a pointy, triangular, arrow shape. Once you get your eye in you can easily tell the difference, but if you're still in doubt, reeds have a burr growing halfway up them. It's very easy to use and needs no kind of processing, you just chop it up like a bunch of spring onions and go. It's hugely popular with my chefs who are always eager for the first sea coriander because as well as having a phenomenal flavour, it's quite an unusual ingredient, being very hard to get hold of.

As you'd expect, the sea herbs and sea vegetables we gather tend to go wonderfully well with fish and crustaceans from the same coast. Andrew Pern at the Star Inn at Harome uses a mixture of North Sea shellfish and sea herbs to make a dish he calls a 'rockpool', which I've had both at his original restaurant and at the Star Inn The Harbour in Whitby.

When we reach the start of 'proper summer' in June, we get into marsh samphire season, a favourite of mine and the chefs I supply. It's known as common glasswort too, because it was once used in glass-making – they burned the plants because the ashes contained the magic ingredient, apparently – and it was used in soap manufacture. But enough of that: in my opinion, the only way you should be using it is as a vegetable with a real taste of the sea that perfectly complements fish and all manner of shellfish, including cockles and mussels. When you pick samphire yourself, fresh off the beaches, there is nothing better. You find it in the culverts first, at the beginning of June, and by the end of the month, it's everywhere. The samphire season is traditionally marked on Midsummer's Day, the 21st of June, and I'm sure you could track that back to the dawn of time. I've had samphire with sea bream, roasted and puréed fennel and bouillabaisse at Mark Birchall's Moor Hall, and with cured halibut, pickled cockles and seaweed dashi at the Star Inn at Harome, in fact I've yet to meet a samphire dish that I haven't liked.

Samphire grows on mudflats and though it's such a delicate-looking, tender green plant, it has the biggest, ugliest roots you can imagine, though when you think about it, that makes perfect sense. Growing on mudflats, it needs to survive the battering of the sea, and those big, fibrous roots hold it fast in the sand and mud however stormy the conditions. The samphire season is quite short, because the plants get really bobbly and woody by the second or third week in September; so one week the samphire is still tender and good to eat, but the next, it's completely gone.

At one point the local east coast buyers for a supermarket chain invited me in to talk about supplying samphire and other sea herbs to them, but I didn't think their ideas for selling them really held water.

'What we could do is put all the sea herbs on the fresh

produce shelf in our local supermarkets on the coast,' one of them said to me.

'I see,' I said, mentally scratching my head. 'So you want to sell samphire to the people of Cleethorpes, where samphire actually grows?'

'Yes. Because it's local, isn't it?'

'Well, yes. That's why anyone who really wants it can go and pick it for themselves. Why would they want to pay for it when they can get it for nothing ten minutes' walk away? In any case, I'm not really sure that Cleethorpes is the right demographic for "exotics" like samphire. Wouldn't you be better trying to sell it somewhere like York where you can't get it locally and where you've got the kind of population who not only know what samphire is, but have the money to buy it?'

I ended up talking us both out of it, but I'm sure it was only a short-term loss because if they had tried to sell samphire, sea aster and sea purslane in Cleethorpes and the neighbouring seaside towns, it would probably have just hung around on the shelves looking increasingly sad, and ended up in the bin. And luckily there's no shortage of restaurants elsewhere wanting to take all the samphire, sea aster and sea purslane I can get my hands on.

We also find sea buckthorn along that coastline, and we used to gather it, because it is a plant with an extraordinary range of uses. Highly salt tolerant, it grows where the sea-spray helps to keep its potential competitors at bay. Its scientific name, *Hippophae rhamnoides* – *hippo* being Greek for horse and *phaos* their word for shining – reflects their ancient belief that feeding its small, bright orange berries to their steeds would improve their condition and make their coats glossy. The berries have been used in Chinese traditional medicine and European folk medicine for centuries, and modern research has established that they're high in anti-oxidants and anti-bacterials. It has even been claimed that they can be effective in slowing the spread of some cancers.

So it's probably not surprising that every now and again sea buckthorn is hailed as 'the new superfood', prompting a flurry of interest which soon seems to fade. It has a very unusual flavour, which some chefs go crazy about, and others hate. I'm with the haters. It's bitter *and* sour. Sour I can deal with, but bitter and sour is a big hurdle to overcome.

To make them edible the berries need to be bletted – exposed to frost – to reduce that astringency. When you press them, the juice separates into three layers – a thick, vivid orange cream on top, a more liquid kind of cream that is very high in saturated and polyunsaturated fats beneath it, and the juice and remaining flesh of the fruit at the bottom. The top two are usually processed into skin creams and other cosmetic products, while the juice and flesh is used to make syrups, jams and wines, or to flavour liqueurs. The orange colour comes from the beta-carotene the berries contain, and while a little is fine, you don't want to swallow too much of it; a few years back, a well-known children's drink was so popular with some kids that the beta-carotene it contained turned their skin orange. I've tried sea buckthorn sorbet and that was pleasant enough, but I can't get over the amount of work it must take to create something edible. I guess that's a chef's job – I'm just glad it's not mine.

Another problem I have with sea buckthorn is that the berries come surrounded by their own armed guards. It encloses its fruit inside an array of spiky thorns that are as long as my fingers and make gorse branches look like tickling sticks. So trying to pick them is horrendous; I'd rather jog naked through a field of barbed wire.

You can actually buy the frozen berries all year round, so if anyone asks me to pick some for them, I usually suggest a specialist London supplier that always seems to have them in the freezer. A chef did ring me up once and say, 'I'm at the seaside with my kids and we've found loads of sea buckthorn. Do you want me to pick you some?'

I heard myself chuckle as I said, 'Yes please, and good luck.'

It wasn't long before he phoned back to tell me they'd only managed to fill half a very small bag before reaching for a family-size box of Band-Aids and going off to find an ice cream van instead.

ON OUR COASTAL MISSIONS, we spend all day collecting ten or fifteen kilos of each plant – as much as we have on order from our chefs and a bit extra for last minute orders – but I'm just as careful about how I cut sea herbs and how much I take as when I'm foraging in the woods. Conservation remains my top priority; I want to be able to come back and find them there next year and all the years that follow. Sea purslane, for instance, grows on bushes, so I'll cut it back by no more than the span of my hand: the top three or four inches. If you just give it a good trim, like the lavender in your garden, it'll come back all the stronger.

It may sound like we're taking a lot of those plants, but the coastal salt marshes are so vast they can be seen clearly in satellite photographs. So there are hundreds and hundreds of square miles of salt marshes around our coasts and whatever you pick from them all grows back within two weeks. The local council were going to bring in contractors to stop the salt marshes expanding ever further on to the beaches, but in fairness, I think between myself, Chris and Ryan, we sort of do the job for them.

Picking those large quantities takes forever, though, and it's all too easy to get completely absorbed in the process, and forget to keep an eye on what's going on around you. Samphire normally grows well beyond the high tide line, closer to the open sea than any of the other plants we gather, or the good stuff does anyway, so you've got to be really careful not to sink into the mud, and to remember that when the tide turns, the sea can flood through the culverts that ribbon their way through the salt marshes, outflanking you before you know it. When you finally look up, you discover that you're already cut off. We all remember those poor cockle pickers who drowned in Morecambe Bay; cockle picking or gathering samphire on the east coast makes you just as vulnerable. It's the reason I always wear short wellies when I'm out there. If you've got your back to the sea, or you're completely focused on the job in hand, as soon as your socks get wet, you know you've got ten minutes to get back to the safety of the shore.

Even at low tide, the retreating tide always leaves a lot of water behind, trapped in the marshes, and you can easily sink into a deep pit hidden by the plants and grasses. Salt marshes are as treacherous as a field that's been trodden into a morass by cows and sheep, which is then disguised by the grass that's grown back over it, and mudflats really aren't flat at all. It isn't like walking on a nice gentle expanse of beach, so keeping your footing is tricky at the best of times.

It would be all too easy to twist or break your ankle, and nobody would know. When you're as far out as Chris gets, you lose your mobile signal, and the marsh grasses hide you from sight. I've told him he should buy himself a flare gun, and that's no joke.

Sometimes you head out in the wake of the tide thinking, 'I've got a three to four hour window and I can get loads done.' You should go to the furthest point of the mudflats first and then work your way back towards the shore, but then you spot a really good patch of sea aster, and find yourself thinking, 'I'll just pick a couple of bags of that right now, because I may miss it later,' and by the time you finally reach the samphire, the tide may already be on the turn.

So coastal foraging has its dangers but I love doing it, not least because you don't half meet some characters when you're out there. There's one old chap, Tom the Cockle Man, who's spent sixty years on that coast and has been picking cockles and clams since his father showed him how to do it when he was a child. He doesn't do it for a living, he just does it because he always has. When his wife was pregnant with their youngest child, he was even out on New Year's Eve, wet through and freezing cold, because she had a craving for clams – now that's true love.

I always collect shellfish to eat myself when I'm out at the coast. Tom once told me that I should pop my clams, cockles and winkles in a bucket of cold water as soon as I get home, and then throw in a handful of flour. The bivalves filter it through their systems, so it plumps them up and gets rid of the grit. In winter the process takes about three days, but in summer, unless you keep your bucket in the fridge, it's only a matter of hours. It's a trick I've shared with my chefs, because it works a treat, producing some of the tastiest bivalves you will ever get your hands on.

Tom has given me a few pointers over the years, but I've also spent enough days on the beach near him to get the general idea. Like me, he comes down only when the tide has just gone out. When I'm ready for a breather and a drink, I'll sit down on a carrier bag to stop my bum getting wet and watch old Tom go about his day, while I feel the sea breeze on my face. As he moves from one spot to another, I can see him looking for air-holes in the sand, or the residue from the bivalves that are buried below the surface, and that's where he starts digging. Tom's an old man now and none of his family have shown any interest in taking over from him, so when he goes, a lifetime's experience and knowledge of the coast and the cockle beds may be lost forever.

There are lots of other characters out at the coast, sea fishing, looking for crabs or digging for clams, razor clams, or sand-worms. We're all entitled to be there of course, though some of them do my head in because they do so much damage. You have to cut samphire, so the roots stay intact and the plant will grow again the next year, so I'll be there, carefully cutting the plants to preserve the roots, but there might be an angler next to me who has a fishing competition coming up that weekend. He'll be digging for the sand-worms that live deep in the sand because, when he sticks one on his hook and casts his line into the sea, the sand-worms are more salt tolerant and so will live – and keep wriggling and attracting fish – a whole ten minutes longer in the water than other kinds of worms, like the earthworms you find in your garden. So he'll be using a big shovel to dig up my foraging spots to get at the worms, and destroying the samphire roots in the process.

Some anglers will spend as much as eight hours digging away there because they are so fanatical about the sea-fishing competitions that they only want worms that will last those extra few minutes. I sometimes think I should be digging them up myself, because they sell for a fortune: you can get about eighty quid just for a bucket of worms. Still, handling worms for a living is not something I would relish.

Even when we can forage undisturbed by anyone else, there is always a risk at any time of year, even at the height of the growing season, that a big storm can play havoc with collecting plants from the shores and marshes. Samphire is like a comb, and a bad storm drives all the seaweed up among it. Then it gets stuck there, and there is nothing worse than trying to clean seaweed out of samphire. It's horrendous, as if you've got a hairbrush that everyone's been using and it's now a mass of tangled, knotted hair and dandruff. The tides will wash the seaweed back out of the samphire eventually, but that takes quite a while, so one big storm can easily wipe out a whole week's work.

Another thing I have to be really careful about is making sure I'm not picking anything illegally. For example, sea kale is a brassica and only indigenous to the British Isles, but it's protected under the Wildlife and Countryside Act and cannot be picked without the permission of the landowner. The Victorians went so mad for it that it was nearly wiped out, and protective measures were brought in. It grows mainly on shingle beaches and can be found all around the country, but it's not that prolific around the coasts that I go to. When I do come across it, I leave it alone. But I heard of a forager who had the police after him because he sent a few vans down to his local beach to clear it out of sea kale. When foragers do things like that it doesn't paint a great picture; obtaining permissions and having some ethics about what we do are vital in the foraging world. In any case, if you do want to eat sea kale, you can buy seeds easily enough and it's a really simple plant to grow. You just put seeds in a bucket with some gravel and a bit of sand and in about six weeks you'll have an abundant supply of sea kale without any fear of the police knocking on your door.

If you want seaweeds, sea lettuce, kelp, red dulse and rock samphire, your best chance is to go where there are lots of rock pools and areas where the seaweed gets trapped, like the Yorkshire coast further north towards Filey and Whitby. We could collect seaweeds as well as the treasures we gather from the mudflats and salt marshes on the coast south of the River Humber, but we can't be all things to all people. Although I understand that seaweeds can be fabulous, versatile wild foods and some of them are full of flavour, a lot don't taste particularly nice and gathering them is a bit of a smelly, slimy business. We already collect over a hundred different plants, and since you can't clone yourself to squeeze more hours into the day and me, Mum and Chris are all knocking on a bit, we've decided not to go up to those North Yorkshire cliffs and rock pools.

Some people love foraging for seaweed. I know a chap in Scotland whose sole task is to go out in his boat and collect the sea kelp that grows in the 'kelp forests' off the coast up there. There is a living to be made from it, and some even hand-dive for kelp like they do for scallops. Good luck to them, but I feel that there are definitely better and more productive ways for me to spend my time – horses for courses.

Nevertheless I am always intrigued by how flavours cross over, and one kind of seaweed caught my attention when I called on Aiden Byrne, who was then chef at 20 Stories, a rooftop restaurant and bar in Manchester's Spinningfields district.

When he saw me, he said, 'Come with me, I've got something to show you. It's a seaweed that smells like truffle.'

That rang a vague bell as he said it, but I couldn't quite place it. We went down two stories to another level where there was a huge walk-in fridge with a carrier bag hidden in the corner. When he opened the bag, it was like a game of 'pass the parcel' at a kids' party, because inside it were three more carrier bags to trap the aroma . . . which is one word for it.

Aiden opened the bag and said, 'What do you make of this?'

It was a ruby-coloured seaweed with tiny fern-like fronds. Then I remembered: pepper dulse, also known as 'the truffle of the sea'. I took one whiff and said, 'Honestly? Not a lot.' It smelt like a truffle someone had dropped on a beach that had got tangled up in seaweed and then gone off. Nature is full of species that mimic other ones, but this was not one of its greatest successes. 'I think I'll stick to the genuine thing,' I said, and funnily enough he was in full agreement.

I've always said to my family that the sea provides and if everything came to a grinding halt, that's where you'd find me, because at least there'd be something to eat. If I was on that TV programme *Hunted*, where the competitors have to survive for two weeks without getting caught, I think I would just get down to a beach and sit it out.

CHAPTER NINE

GOOD ENOUGH TO EAT?

CHRIS AND I COLLECT wild plants all year round and from all sorts of different places: fields, woods, river banks, marshes, peat bogs and moors, mudflats, sandbanks, rock pools and coastal cliffs. It's a bit like a giant supermarket except that the aisles are a bit further apart and fortunately, unlike Tesco's, you can always find the same thing in the same place. We pick many of the same plants year after year – truffles, puffballs and mushrooms in general have become some of my staples. Just as a chef or restaurant, or even a waiter, is only as good as the last meal they've served, I know that my professional relationship and reputation depends on maintaining the quality and reliability of the wild foods we supply.

Any spare time – and there's rarely much of that – is spent constantly searching for new ones we can add to our range and for different ways to use the familiar ones. Having found and gathered them, we deliver them to our chefs and, when they haven't previously used a wild ingredient, we explain its history, flavour, profile and its culinary or medicinal properties, how we came to discover it and learn the best way to prepare and preserve it, and even how other chefs have used it in their dishes.

*

Everything we do is planned down to the last detail, including the ebb and flow of the tides when we're picking sea herbs, because the chefs want deliveries by Thursday at the very latest, giving them time to prepare everything for their weekend peak periods. They usually give us a bit of warning to get our picking list together but to be honest, it can be like trying to herd cats. If you send a chef a text, it might take him a week to get back to you, but in his mind it has been an instant response. It took me a while to get that but I'm quite used to it now. Chris isn't, it drives him nuts. Even after all these years, I still get it in the neck when they order a ton of something, but to be fair, these days it's usually Chris who actually has to go out and pick it, and though he may chunter about it, he always does a great job.

Despite sometimes taking an eternity to respond, chefs still tend to want everything immediately and preferably out of season as well. It's not unknown for us to be asked for meadowsweet in January or wild garlic shoots in September, though sometimes we are able to oblige them, because if you know where to look, quirks in local soil conditions and microclimates mean you can find pockets of plants that are growing well out of their normal season.

Nothing I gather and supply to my chefs has been processed or stored – I don't even possess a cold store. I send out a weekly list, telling the chefs what is already in season and what is about to be, then we go out and pick the plants and deliver them the same day or the day after. We pack everything carefully and surround it with ice to keep it fresh.

When we bring back sea herbs and coastal plants like samphire, they're often covered in mud. Most people would just wash that off straight away, but if you maintain the things you've picked in as close to their natural condition as possible, they'll last much longer. So we keep the things that we've gathered in the mud we found them in until they're ready to be used, which stops the acids escaping and preserves them much better. If left in the mud, they'll last for two or even three weeks, whereas if you wash them, they'll go off much quicker than that. Squeaky clean is not always best for preserving things – just think of those 'bog bodies' that have been preserved in peat bogs and deep mud for thousands of years. I'm not saying that samphire will keep that long, but the bit of natural muck that it comes with certainly does it no harm.

I'M ABSOLUTELY meticulous about where things are picked, how they're picked, what they've been growing on – every single thing about them, in fact. If one of my chefs asks, 'Where have these chestnuts come from today?' I can tell them exactly which wood and exactly which tree, and can pinpoint the location on a map down to five square feet. Hyper-vigilance and being a bit of a pedant can be a good thing sometimes, and fortunately all the members of my family have great patience when it comes to sorting out our produce. Whether it's bilberries or sea aster, as long as we have something to watch on the telly – like Mum's favourites, *Xena: Warrior Princess* or *The Walking Dead* – you'll find us with bowls of bilberries on our knees, wielding tweezers and magnifying glasses to pick out all the leaves and stalks.

One of the ways I ensure the quality of what I gather is by making sure the plants have grown in an environment that is free of toxins – especially mushrooms, which are like sponges, and will take toxins from the soil, along with the nutrients they need. If you pick them from somewhere that was once a brick pit, a tannery or an engineering works, you have to be mindful of the possibility that there's a cocktail of toxins or heavy metals in the soil, and you really don't want to find yourself swallowing lead, mercury or goodness knows what else with your dinner.

French vintners talk about *terroir*, where even a slight difference in bedrock, soil, aspect, altitude, drainage and so on can make a significant difference to the flavour and quality of their grapes and the wine they make from them. If they can nail it down that finely, the same rules must apply to every plant in every location. I'll never forget doing some load lifting at an airbase in Italy while I was in the Navy, and finding plum tomatoes springing from the cracks in the edge of the runway. You couldn't imagine a worse environment for them to thrive than that arid, baking concrete, and yet they were the best I've ever tasted. A random seed must have taken root there, most probably deposited in a bit of bird poo, and that precise spot happened to have everything the tomatoes needed. Nature has a way of deciding its own place in the world and, as any gardener pulling weeds will confirm, you'll never stop plants from growing in places where they do well. Fortunately this airfield didn't see much action – it was about three square miles with a plane arriving about once every week, so there wasn't much chance of any aviation fuel seeping through.

If you're driving along a motorway in spring, you'll often see thousands of tiny white flowers growing along the verges, the same scurvy grass plants that I gather from remote, sea-washed beaches. These salt-tolerant plants thrive on roadsides because of the saline grit spread to stop us skidding out of control in winter. If I picked them, they would taste the same as those I pick on the east coast, but all that lead from car and lorry exhausts means you definitely wouldn't want to eat them.

While some potentially edible plants are already full of toxins or acquire them from the soil, other inedible ones can benefit us by purifying the air or water, filtering out the toxins, just as a garden hedge in the city will filter the dust and pollution before it reaches your front door, or as reed beds in India counteract dysentery and other water-borne diseases.

Other plants can give you clues about what poisons the soil may contain. There are old smelt mills and spoil heaps in the Yorkshire Dales on which almost nothing will grow, even though it's well over a century since the last lead mine closed down, and yet you will almost always find those beautiful little wild tricoloured pansies known as heartsease growing there, because they have such a high tolerance to lead that they can thrive where other plants will die. So places where you find a single dominant plant are telling you something – in the case of heartsease, that the surroundings are very high in lead – so I steer clear.

I just can't afford to supply a chef with something from a contaminated environment. When I was on a course to get the food safety and hygiene certificate I needed to supply the Christmas Tree Syrup to Lakeland, the instructor told us about a death from salmonella poisoning that had been traced back to a restaurant that had served a parsley garnish containing faecal matter. The risk of contamination makes me doubly meticulous about checking sewage reports and so on when I am working on the coast, and I even raise my safe picking height by a good twelve inches if I see a dog stand on its front two legs and pee up a tree backwards. But I'm collecting wild plants, not cultivated ones, so I can't give a written guarantee that a bird hasn't crapped on them or a badger hasn't run through them.

WHEN YOU'RE collecting plants it's not just about where you're picking them, it's also about when. You always have to be thinking about the time of day: whether it's cloudy or sunny and whether the bugs have warmed up sufficiently to be looking for breakfast.

We always pick fruits in the early morning or evening to keep them cool, which minimises the deterioration of their sweetness and stops their aromatics being released to the day. Mushrooms deteriorate even more quickly after they have been picked, so we gather them in the cool of the evening, then clean and prep them, and send them out for delivery first thing the next morning. On the other hand, it doesn't matter what time of day we gather nuts because their shells protect them and preserve their flavour for much longer.

When it's spring it's all about the light; when it's summer it's all about bug avoidance. This helps us not only to preserve the plants we're gathering and keep them in peak condition, but to preserve ourselves too, by avoiding too much discomfort. I try not to go out in the midday heat, not only because plants deteriorate quicker if picked then but because it's the time when bugs, flies and biting insects tend to be at their worst. Bugs don't like the cold, so we can generally avoid the worst of them by picking before eleven in the morning or after four in the afternoon, though clouds of midges can be a problem on summer evenings. If you stand up after being bent down picking something, you often find you've stuck your head right in the middle of a midge swarm. The little swines are immediately in your eyes, ears, mouth and nose as well as busily burying in your hair to bite your head.

Insect bites are an occupational hazard. Midges are the most common; the mouth parts of a midge act like a razor saw and they leave the nastiest of bites. Mosquitoes aren't great either, one bite can result in about two weeks of sore itching. There are even certain spiders that bite, and although the venom isn't strong enough to do any damage, it's still a shock when it happens. But it's horse flies that are the worst; being bitten by a horse fly is like being injected with a blunt needle it's so painful going in. And then the swelling is the size of a golf ball.

IT IS OBVIOUSLY essential that I correctly identify the ingredients I am supplying my chefs. Many plants look very similar, so in my experience it's always better to use a few different foraging books to cross reference the information you're looking for. For example, many sing the praises of the first shoots of young bracken – the tightly curled fronds that earned them the local nickname of 'fiddlestick heads' in Yorkshire, because they look like the scroll at the top of a violin neck. They are regularly eaten in Canada and North America, apparently perfectly safely, and are sometimes flown over here when in season. Although British bracken looks and tastes exactly the same, if you try to eat it without considerable processing, you'll give yourself a very upset stomach. All I can do is warn of the dangers, but some chefs are willing to risk it anyway. Mushroom Martin has many years of foraging experience behind him, and told me the North American variety has a different DNA structure. I guess they wouldn't import it if that wasn't the case, because the countryside here in the UK is absolutely chock-a-block with the stuff.

I often get asked about mushrooms. Mushrooms and fungi really are a vast subject to which you could dedicate your entire life; they are fundamental to the functioning of the whole planet. From my point of view though, I just point out that roughly two per cent are deadly poisonous and about ten per cent are wonderful and edible. The rest you don't really need to worry about. There are loads of edible mushrooms in this country but I am determined to only supply my chefs with mushrooms I'm totally sure of identifying, as we seem to be very good at growing poisonous ones too. I always have to remind people that every mushroom is a little bag of chemicals, so you can't allow yourself to be fooled by their appearance and must always be a hundred per cent sure of your identification before you even think about eating them because an edible mushroom may have a dangerous doppelgänger. When fully open, parasol mushrooms look like a beach umbrella, and they are good to eat, but they can be confused with the deadly panther cap, one of the amanita family of poisonous mushrooms as the only difference is that the parasol mushroom has a white stalk with brown flecks on it, while the poisonous panther cap has a plain white stalk. So, although I have tried them myself, I won't pass them on or sell them to other people, just in case I make a mistake in identifying them. Even if I'm ninety-nine per cent sure, it's still not something I ever want to risk.

My grandad ate a lot of mushrooms while he and my Uncle Ted were living off the land in the Polish forests, and although mushrooms were and are much more prolific in

Poland because of their profusion of ancient forests, there are 4,500 kinds of fungi in the UK. I only pick and sell about fifteen of them because even if I'm happy to eat more myself, I just can't afford to take the risk when I'm supplying them for other people. A lot are LBMs – Little Brown Mushrooms – and they vary wildly in toxicity. The tiniest can be lethal. The chestnut dapperling, for instance, sounds charming and is immensely pretty, but it's a real killer.

Some of the poisonous ones are so beautiful they really do look good enough to eat, like scarlet elf cups, which look magnificent, cup-shaped and as vivid as a poppy field. They grow on rotting branches among the leaf litter on the forest floor, and in traditional medicine, their flesh, dried and ground, was supposed to have styptic properties – stopping blood flow. However, they burned my fingers when I picked some, so I thought, 'I'm definitely not selling them to my chefs; if they burn your fingers when you're trying to pick them, heaven knows what they'll do to your mouth if you try to eat them.' I know some will disagree with me but that's just a personal preference.

Chefs still ask me for them, but I always refuse. I'm not risking it. I have two mushroom identification books – one shows a skull and crossbones alongside the scarlet elf cup and the other shows a knife and fork, but for me the scarlet elf cup definitely belongs in the first group. So your motto should always be: 'If in any doubt, don't pick it and don't eat it.'

Some mushrooms are so toxic that I wouldn't even touch them, never mind eat them. Death caps, for example, can be whiteish when they're young, a bit like field mushrooms, but they grow out of a sac like a lot of other poisonous mushrooms, and if you are foolish enough to eat one, you're likely to find out why they got their name. The *Horse Whisperer* author, Nicholas Evans, despite being a regular eater of foraged food, nearly killed himself, his wife and two friends by mistaking highly poisonous deadly webcaps for girolle mushrooms. He became so ill that he would have died had his daughter not donated one of her kidneys for a transplant for him.

Even worse is the poetically named destroying angel. It's a pure white mushroom and disarmingly safe-looking, but you need to be very alert for it because if you're foraging in a big patch of field mushrooms and there's a destroying angel among them, you'll be toast. Even a piece the size of your little fingernail will dissolve you from the inside out. And once you've eaten it there is no antidote, no cure, that's it – you're a goner. If you get yourself to hospital, they'll just make you comfortable and call the undertaker. It really is that serious. The symptoms begin a few hours after you've eaten, with stomach cramps, vomiting and diarrhoea. That phase can last a few days and may be followed by what appears to be a complete recovery, but I'm afraid that's a cruel false dawn. Alpha-amanatin, the main poison this dark angel contains, will kill the liver cells and then pass through the kidneys before being recirculated through the body again in a vicious circle that has only one end: complete kidney and liver failure.

In 2019, twenty-eight customers at a Spanish restaurant suffered from severe food poisoning, one of whom died. It was suspected the culprit was a dish of morels that had either been incorrectly identified or not properly prepared. All morels contain a toxin, hydrazine, that affects the central nervous system. In the old days they were steeped in Madeira wine, which gave them a fabulous flavour, and leached the water-soluble toxin from them. So without careful preparation they are dangerous to eat, but are such a frequently used ingredient these days that I fear complacency can sometimes creep in.

And it's not just mushrooms that you need to be careful about either. Some of the most beautiful berries absolutely must not be eaten. One such berry is the berry of the spindle tree. These trees are quite thin and sparse, and not that common, though we do have a row of them in the woods by our house. The wood of the spindle tree, as the name suggests, was used in weaving, as well as for clothes-pegs, knitting needles and toothpicks. In winter they have the most beautiful fuchsia pink berries with bright orange seeds inside them and really stand out when the leaves have dropped off and all that is left are the vibrant pink fruits against the background of the dark woods. They look absolutely fabulous, good enough to eat, in fact, but if you do, your next stop will be A&E, because they're poisonous. Their only use is medicinal – in the past the berries were dried, powdered and used as a treatment for head lice on humans and mange on cattle.

Animals don't tend to make the same mistakes as humans. Evolution has taught them what they can eat and what they should avoid, and even domestic dogs and cats have enough of the wild genes of their ancestors to know what's safe. I am never bothered when Fred is out mushrooming with us, because he is an expert at stepping around them – he does a little tiptoe mushroom dance to avoid them – and is totally disinterested in them. I have to say dogs surprise me with what they will and won't eat. On the one hand some will eat a six-foot length of Christmas tinsel and a dustpan and brush, while others will know to skirt around nature avoiding the deadliest of plants. Dogs' noses are so much more sensitive than we give them credit for, and clearly alert to dangerous chemical compositions. People often get in touch with me in autumn because they've found what they think might be poisonous mushrooms in their garden and are terrified that their dog is going to eat them. 'How can I get rid of them?' they ask, and I simply tell them, 'My Fred will eat a maggot-ridden mole that's been dead for six weeks, but he won't touch a mushroom. But get rid of them by all means, some dogs *are* daft enough to eat them. But on the whole, considering the amount of mushrooms that grow, it's a rare occurrence.'

ONE OF THE REASONS chefs like foraged wild food is because it hasn't been forced into production. The natural response of any plant to being cut or processed – and it's true of everything from tomatoes to fresh green herbs – is the reason food always tastes better if you cook with fresh ingredients. You're getting the best possible flavour at that point in time, whereas if you cook with processed food, all the active ingredients tend to have died a death a long time ago.

When something grows in the wild, it's overcome all the potential obstacles and adapted to a very specific environment, but even experienced gardeners may struggle to cultivate that same plant in a domestic situation. You've got to find or create the right soil conditions, the correct pH balance, the right amount of nutrients and sand and grit for drainage, choose the most suitable aspect and the right amount of exposure to the sun – full sun, dappled shade or deep shade – and the right protection from wind, frost and snow.

Having done all that, you still have to defeat all the pests and diseases that prey on it. There are acres of books and newspaper columns dedicated to these myriad challenges: why are my carrots not growing? why have my potatoes got blight? why have my roses got rust on them? why has everything died this year? There's always something in a garden, the list of problems is endless.

Everything in a garden tends to be full of bugs and afflictions – greenfly, blackfly, mildew, rust, blackspot, etc., etc. – and you've got the spray gun going full time. Yet there are few such problems out there in the wild; plants either grow or they don't, and things seem to survive perfectly well without anyone running round picking off the aphids or covering them with chemicals. It all comes back down to Nature's capacity to overcome all potential obstacles, allowing her fruits to thrive in the ideal spots they've found for themselves.

Scientists modify plants in order to increase their yield or their disease resistance, but pathogens and viruses are constantly evolving and developing – so if you're developing a rice that's resistant to x, y, z diseases now and replace the vast range of traditional varieties with it, if a freshly mutated pathogen or virus then comes along and destroys it, what on earth do we have to fall back on? We might have stored seeds in underground seed banks, but that wouldn't be enough to combat a major crop failure overnight.

I see how plants survive and flourish on a daily basis, so I have to admit that genetic modification terrifies me. The idea that science can 'improve' on Nature isn't one I subscribe to. I understand that overpopulation and food poverty will only get worse as time goes on, and we have to do something in order to produce the sustenance we need to survive, but why can't we try to recreate the best possible growing conditions in the wild, rather than manufacture a new breed of robo-crops?

Scientists can only deal with what's in front of them, they can't accurately predict what will occur in the future. When I'm out and about, day in, day out, I'm seeing plants that have adapted to their particular environment and conditions over hundreds, thousands, and perhaps hundreds of thousands of years. The moment we come along and start interfering with that, we really don't know what disasters we're storing up for the future.

CHAPTER TEN

MICHELIN STARS

ALMOST ALL THE WILD FOODS that I forage are destined for the kitchens of top chefs, most of them running Michelin-starred restaurants. I was looking at the new additions to the Michelin list recently and almost all of them had menus that used foraged ingredients and were seasonal. I get a lot of pleasure out of that, because nothing is as good as natural, and the list of chefs, past and present, that I've supplied reads like a bit of a *Who's Who* of fine dining.

In total, the chefs I currently supply have twenty-odd Michelin stars between them, and that total is being added to all the time, but I supply a lot of other chefs who don't have Michelin stars – yet – and they are just as important to me. I have a golden rule only to work with people I like, and each and every one of them is not just a customer but a friend and supporter, always as happy to pool knowledge and share some of their secrets as I am.

They can be very generous too. I asked one if he could do some special cakes for a fancy afternoon tea I was hosting at home for my uncle's birthday.

'How many do you need?'

I thought, 'Well, there's only me, Mum, Chris and Robert,' so I said, 'Eight will do it.'

'OK, leave it with me.'

When I went back to pick them up, out came two giant trays of the most amazing creations you could ever see. To my embarrassment, he'd thought I meant eight guests, and because he's a very generous man and wanted me to be able to impress them, he'd made them eight cakes each – sixty-four altogether. I didn't have the heart to tell him that might be overdoing things a bit, so after my uncle had left with a box to eat at home, I spent the rest of the day distributing the leftovers to my neighbours.

Anyone can become a chef or work in the hospitality industry; it's one of the most inclusive trades there is. Whether you're pink, black or sky blue, fat or thin, with the IQ of a gnat or an Einstein, it really makes no odds. If you embrace people and food, you're in, but that's when the hard part begins and the road to respect can be a long one, because, like all professions, you have to build up a body of work.

Being a chef combines technique, experience, science, chemistry, alchemy and creativity, with a generous dash of genius on top. A chef has to be a good team leader and business-savvy, to be able to work a hundred hours a week and keep those varicose veins under control. Some are very financially astute, with one eye permanently on the bottom line and making maximum use of every ingredient, but others are less concerned with money and more interested in the dish itself. One chef I know will cut a whole fillet of prime beef into neat squares and then chuck the rest in the bin. My dog has lived very well off the things chefs like him have thrown away, but if I'd caught the fillet before it hit the bin, it would have been me having Chateaubriand for tea and not the dog. However, the same chef willingly gives his time up to cook for charity, so he's very far from being a prima donna.

THOSE CHEFS HAVE more culinary flair and skills than I'll ever have, and are right at the very top of their game. As well as great technical skills, chefs need to know from their ingredients bank what might go with what and a palate that tells them when something is as good as it gets. No one is booking months ahead to go and eat at chain restaurants, but people will wait months and give their eye-teeth to eat at some of the Michelin-starred establishments that I supply. It takes talent, hard work, ambition, and years of training, experimentation and experience for a chef to get to that very top level. The majority will never make it to the echelons of Michelin, but those that do deserve to be there. It's an accolade that can't be bought, bargained for or anticipated and that's why it is so respected by chefs. A Michelin star can also mean an extra £100,000 turnover a year, and that brings an immense pressure as a chef slaves to maintain the star, which can be brutal.

Michelin's list is constantly reviewed, and sometimes restaurants drop off but then reappear again a couple of years later. This can depend on what their head chef is interested in or learning at the time. Chefs are always experimenting and sometimes it takes a while to get a new menu up to the level of a Michelin star. I have one chef who is currently trying out a Polish and South African fusion – you couldn't even make that up. Every chef has their own understanding of how flavours work and I see that all the time with the chefs who I supply. Each of them could be given the same ingredients and come up with something totally unique and delicious.

The best chefs have right-hand people who are integral to keeping up those standards, it's always a team effort. Mark Birchall was the right-hand man at Simon Rogan's L'Enclume for years, helping to maintain the restaurant's two-star standard. He's now supported by the brilliant James Lovatt at his own restaurant, Moor Hall. Similarly, Stephen Smith, who is director of Andrew Pern's Star Inn at Harome, is a phenomenally brilliant chef in his own right.

WHILE THE CHEFS are in the kitchen perfecting their craft, I'm out in the fields and woods perfecting mine. It's not just a case of picking the plants and dropping them off at the restaurant's back door; when chefs buy something from me, they're investing in a lifetime's knowledge and expertise. They need to know what a plant's particular characteristics are: where it came from, how long it will last, what processes it requires, how to preserve it and how to extract the maximum flavour from it. I also need to be able to tell them when it's going to be in season and for how long, when it will be absolutely at its peak and how often I can supply it, so that they can plan their menus around it.

They already have a huge amount to think about in order to be able to create those show-stopping dishes. So if I can share a bit of off-track information about what I'm supplying, that removes a little bit of their pain and pressure. They don't then have to spend hours, days or weeks experimenting with different preparation, preserving and cooking techniques to bring the best out of them, and they can focus instead on creating a fabulous dish to showcase them. Sometimes I come up with ideas of my own that I'm sure are going to work, but I have no idea how to execute them. So I'll describe my idea to a chef and they can work out the technicalities of how to bring that idea to fruition.

The chefs need to know the potential pitfalls of any wild ingredient they're going to use and I try to really lay it on the line for them. I might wax lyrical about the qualities of one plant but in the same breath I'll be describing another as 'minging and manky, but to extract the best flavour from it, this is what you need to do . . .' I find it's best not to metaphorically sugar-coat things with chefs. I don't do a hard sell. If something is tricky to process, I will tell them. If it's a pain in the backside, I will tell them. But sometimes the harder an ingredient is to deal with or to create something out of, the more the chef will warm to it. I guess some people like a challenge.

I've learned how to bring things together, and prepare and preserve wild plants to make something really great from them. I love being able to share that knowledge with chefs to inspire their creativity. It's a symbiotic relationship; when the two halves come together it can be phenomenal, and I get as excited as them about finding a new wild ingredient and discovering its history.

Sometimes I'll put a plant in front of a chef and he hasn't got a Scooby Doo what to do with it. I can help him work out how he can make something fabulous with it; I might even suggest possible flavour combinations. For example, chefs love to pair grouse with the ingredients they thrived on. A lot of them use a sprig of heather as a garnish, but they really shouldn't, because although it has traditionally been used in beer-making and infusions, heather is poisonous. The only part that you can safely eat is the pollen it produces, and that does have a lot of flavour.

We get a lot of bell heather – the original British wild heather – up on the moors. When it's in full bloom, we get a big plastic bag, hold it under each plant, give it a shake and collect the cloud of pollen from it. Mum and I might even be seen giving the heather a good kicking, but it's quite robust and can take a beating. We're a long way from Wordsworth wandering like a cloud through a host of daffodils. It takes forever to collect and it's so light that it's like trying to weigh air, but when it comes to the Glorious Twelfth of August and the first grouse of the season have been shot and rushed down to a restaurant to be cooked for the guns' supper that night – they should really be hung for some time first to improve the flavour, but the diners want them straight away – there is nothing better to baste them with than heather pollen butter.

If you can't get the heather pollen, a Douglas-fir-infused butter complements the grouse well too. They are then often served with bilberries and lingonberries on the side, and perhaps a few orange birch boletes mushrooms, which grow on the stunted moorland trees. So you've got the grouse's circle of life right there on the plate. It all makes for an expensive dish, but if you've spent a few thousand quid on a day's shooting, I guess a wonderful dish cooked by a top chef is just the icing on the cake.

Having made sure they're fully in the picture, when I bring something new to my chefs, I can see their excitement, particularly if the plant is growing locally and has a whole history behind it. The chefs in turn enthuse their servers, who pass on that excitement to the diners. It's such an important part of creating a really special moment for them.

It's always a thrill for me to look at restaurants' websites and see things that I have brought them featuring strongly on their chefs' menus. I don't care if it's the main ingredient for a dish or a bit of frippery on the side, it's always nice to see. I just wish the rest of my family didn't have such plain tastes. They're happy to make things at home with wild food but they don't want to go out to 'fancy pants' restaurants because it's too much of a palaver and they're worried they'll miss something on the telly while we're out. There is always always football on the TV, which rules Chris out, and with my Mum it's usually something sci-fi, although I have prised her out a couple of times recently and she's enjoyed it.

I've eaten more than my fair share of stunning food prepared by top chefs and I'm always willing to give unusual dishes a try. I've had a go at everything from wood ant bums to mustard-seed ice cream with a horseradish tuile to Mangalitsa pig fat. Mangalitsa pigs are very hairy and frankly I'd rather have one as a pet. Even if the fat is cured for weeks, it's still just a lump of raw pig fat. I try to blank out what the waiter is spending five minutes explaining, and just eat the dish. It can be utterly surprising and delightful, and I'd be even happier eating it if the waiter would just stop talking about how it's made from a retired dairy cow that was sent to slaughter after a lifetime giving milk. I'd be a vegetarian if I liked vegetables enough, but I have pointy teeth for a reason and retired dairy cow is actually quite delicious – but I do want something to have had a good life before it hits my plate.

Without exception, all the chefs I work with are incredibly generous and love to have a captive audience at their table. I often have to debate with myself whether to accept an invitation for lunch or dinner because I know it could be never ending. A cup of tea and a sandwich can often turn out to be a full-blown affair and I often haven't got the time to sit down and have a meal, even though I know that they really want me to try the food they have created.

A few months ago, I finally had lunch with the wonderful chef John Feeney. I lost count after twelve courses and three desserts. But, dear Lord, it was phenomenal and I was very apologetic for not having accepted his invitation sooner. It is often the case.

I also finally visited Joro in Sheffield, a place I have supplied for a few years. The restaurant – which is in a building made of shipping containers – and its chef owner, Luke French, have a phenomenal reputation. But, despite being only twenty-five minutes down the road, to my shame I had never eaten there before a couple of weeks ago. It was a stand-out meal for me, and will be a matter of months – not years – before I go back.

When I have the time, I am always excited to sit down and discuss ingredients and menu ideas with chefs. I am flattered that they are happy to bounce their thoughts and ideas around with me and usually we will both come out the other end inspired by the other.

CHAPTER ELEVEN

EXPERIMENTS WITH FLAVOUR

IN BUILDING MY CAREER as a forager, I've been able to draw on the knowledge accumulated by previous generations of my family and the research of scores of scientists, botanists and landscape historians.

I will draw on any source I can find, from 'proper' science to primitive medicine, folklore and oral tradition, and there's often more than a grain of truth in the common names of plants and folklore and country sayings. A lot of plant lore and wisdom has been lost over the years, but some of it is now being rediscovered. Even the things that many dismiss as myth and fairytale are often worth exploring, if only to give you a starting point. Nature often tells you what's going on – it can be staring you in the face if you only take the time to look – but we can be so preoccupied with checking our phones, or finding out what's been happening on the telly that we don't see what's going on in the real world right under our noses.

An interesting example is the yew tree, which is a common sight in churchyards all round the country. Folklore tells us that yew trees have 'one foot in heaven and one foot in hell', and you have to ask yourself why. It may simply have been because they grew both in the wild and in church- and grave-yards, but perhaps the foot in hell refers to the fact that every single bit of the yew is toxic, with the exception of its berries – provided you don't eat the pip inside them, which is really poisonous. I often wonder how many people got very ill or died before making that discovery.

The yew tree produces a compound that, once processed, is now used in cancer treatment. An extract from the yew needles, which have to be from the previous year's growth, is used to make the chemotherapy drug docetaxel, also known as Taxotere, so perhaps the foot in heaven is an acknowledgement that there was once a folk remedy using yew needles, though the knowledge of what it was and how to use it has since been lost.

My developing knowledge of plant history and folklore has led me to many other edible and medicinal plants. I think now more than ever we need to rediscover the medicinal or dietary uses of plants that have been used for centuries in traditional remedies. Whenever I find myself having a quick chat or doing a talk about foraging to a roomful of people, they're always interested in the wild plants I gather and their rich variety of uses. I don't think I've ever spoken to anybody who hasn't asked for more.

Don't get me wrong, if I was diagnosed with cancer, I wouldn't be going out and sucking the stem of a root to try and cure it. I'd be straight off to the hospital and getting my fill of conventional medicine, chemotherapy, radiotherapy – the works. But I would also eat puffballs, because they contain a compound called calvacin that has produced a really effective anti-tumour drug, and there are plenty of other traditional remedies that I can – and do – make use of. Calvacin is quite fascinating really – there are plenty of medical research papers evaluating it, and on the face of it they had discovered an anti-tumour drug with huge potential. However, they were unable to cultivate the mushroom itself to go into a viable economic production of the compound, but I hope someone revisits it one day because technology has moved on quite a bit and I am convinced that the mushrooms themselves can be cultivated.

Common names of plants and their usage go hand in hand, of course. Some are blindingly obvious, and do exactly what it says on the tin. Fat hen makes your chickens fat, though humans can eat it too. Fat hen has a leaf like a goose-foot, is very high in vitamin C and also tastes a bit like spinach. It's a prolific weed in gardens and crop fields and was widely eaten as a vegetable from Neolithic times through to the end of the Middle Ages, when it was largely replaced in people's diets by plants like cabbage and spinach. Scurvy grass, packed with vitamin C, has been eaten by sailors since medieval times.

Sometimes a plant's common name tells you as much about its effects as its uses. Water pepper – *Persicaria hydropiper* – has another common name, and it's very common indeed: 'arse smart'! Its leaves taste as hot as chillies or wasabi, but back in medieval days, rather than eating it, people commonly put it in their bedding because they reasoned that since it was hot and an irritant, it would get rid of their fleas and bedbugs. It didn't, of course – it just made their bum sting instead. So far I've never managed to get my hands on any, though I did catch a glimpse of some once, growing in puddles in the central reservation of a motorway, so I still have hopes of finding it one day without having to risk life and limb. The head on it reminds me a bit of those liquorice sticks that have hundreds and thousands on the end. It's quite prolific in some parts of the country, just not my bit; either that, or I've missed it, which would be annoying but not unthinkable. I've been told by other foragers that it likes to accumulate around permanent puddles, but I've been puddling all over the place and I've still never found it.

Some of the plants I forage have surprising flavours, such as gorse that tastes of coconut. Mayweed is another plant with a very unexpected flavour. There are several varieties, including scented mayweed and the rather less appetising stinking mayweed, but pineapple mayweed takes top flavour prize. You often find it growing in the cracks in pavements and in areas of dry sandy soil, and if you pick some, rub it in your hands to release the natural oils and then taste it, trust me, you're more than half way to your Caribbean paradise, the pineapple flavour and smell are amazing. Lots of chefs crystallise mayweed as a garnish, or you can infuse it in alcohol to make a vodka with something of a tropical taste. It's better than you can imagine. However, with it being a pavement-, track- and footpath-dwelling plant, you do have to be careful where you pick it and have to find a place where nothing and nobody really ventures. Fortunately a friend of mine has quite a bit of land that doesn't have public access, so I can usually pick it from there with impunity, confident that nothing has really contaminated it.

Discoveries like these really got me thinking about the science of taste. If you can find the same distinctive flavour and aroma from both a nut and a flower, and from a fruit and a weed, they must have a compound in common, or at least one that is very similar. Nature is ruled by formulas and patterns and I think these can be broken down into more than just species. So I'm currently trying to identify and isolate the compounds in every one of the plants we gather and establish whether they're oil, water or alcohol based. It's no easy task, but the rewards are considerable. For instance, wonderful, aromatic sweet cicely lasts for a good couple of weeks in the fridge, because it won't break down and release its flavour in water. Sweet woodruff, on the other hand, soaks up the moisture in the cool-box but then disintegrates in two days. So I'm charting all the compounds and how they break down, which I think will be a useful way for chefs to look at botanicals, and perhaps help them find more unexpected combinations of complementary ingredients in a more scientific way.

The best chefs are always pushing things to the next level, with fermentation studios and the like, to the point where their development kitchens can look more like chemistry labs than food preparation areas, so I can never rest on my laurels – I need to build my own knowledge of how to get the most out of the wild plants I collect, so that I can match my customers' ambitions.

Proper botanists and biologists are studying many of those plants too, but focusing on their chemical properties, not their taste, whereas I can see patterns forming and connect the dots in different ways. I'll be looking at groups of tastes like aniseed in certain mushrooms and plants. It's a bit nerdy but quite fascinating, and gives new meaning to the creation of a flavour profile. Flavours are truly global. It blows my mind to know that you can find plants that taste like pineapple or coconut on the opposite side of the globe and in completely different manifestations.

There are botanical societies in almost every county, whose members have an encyclopedic knowledge of local flora and fauna. A lot of societies do detailed reports on certain areas and publish the findings, which is a godsend, making my job a lot easier when I'm on the hunt for a particular plant or tree, as the reports will identify the genus and specify the locations where it's been spotted. But knowing the science and likely location is only part of it. I love discovering the stories of the treasures I gather. There are still people who have a fund of country lore and knowledge about the traditional uses of wild plants that their parents and grandparents taught them, but unless they're able to pass on their knowledge, it will die with them. I've often had the good fortune to bump into such people while I've been out and about on the salt marshes, in the woods, walking through meadows, rambling over moorland or striding up hills. I love to spend time with them and listen to their tales, and luckily I've got a memory like a sponge so everything they tell me, every story behind every leaf, shoot and bud, is in there somewhere, waiting for the right moment to make itself useful.

I've also carried out loads of my own research. My scientific knowledge is entirely self-taught but, strangely enough, my degree has helped me no end. I've always been interested in patterns. Patterns of human behaviour inform the law so learning law is not a million miles away from learning maths – you just have to digest the formulas.

So there are formulas to everything, like the way certain colours work together in a painting, and with a keen eye, you can spot patterns in nature too. Suddenly, even the best disguised plants become glaringly obvious. You get the imprint – the pattern of what you're searching for in your mind's eye, whether it's a minute variation of colour, leaf shape or berry – and as soon as you have it, you see it everywhere. It was like that for me with lingonberries. When I first found them, it was really exciting as I knew instantly what they were when I saw them and yet realised I must have passed them scores of times before, when I was looking for other things, without ever noticing them. I could kick myself sometimes for things that I've missed.

I also look for patterns in the storage of the plants I pick, puzzling over why some stay fresh and others fade. It may be temperature, it may be the presence or lack of circulating air – put some mushrooms in plastic bags and they'll 'sweat' and deteriorate fast, whereas others will stay in good condition – it may be a need to be kept in water, in oil, or in alcohol, or a host of other factors. Learning how a plant flourishes in its natural environment, and why, also helps; if I can replicate those conditions, it may keep for much longer.

Much of it is trial and error, but I'm a big believer in doing the research. I have to because I need to know, for example, whether something like the phototoxic sap in hogweed that turns to acid and burns you in sunlight behaves in the same way under a strong artificial light in a restaurant kitchen, or is safe to handle and eat. The answer in that case turned out to be that it is still toxic, but cooking breaks down the chemical compounds and brings out its lovely aniseed flavour.

So a lot of my work is discovering how the aromatics and compounds from the wild plants I pick can be combined to enhance their taste, and I've used some of that knowledge to create gins from the herbs and botanicals I foraged. If you set up a whisky distillery, it has to be in a barrel for six years before you can sell it, and then you have to leap through a whole load of additional hoops created by Her Majesty's Revenue and Customs, so it's an arduous and expensive business. Making gin is much more straightforward, though it's still not cheap to do – starting a gin distillery from scratch is still going to set you back about £65,000 unless you're a farmer and can get a change-of-use grant. Juniper is the only requirement for gin, that's all there is to it. You can make 'bathtub' gin, by simply adding a flavouring to a bought-in spirit, or if you're feeling more ambitious, you need to either vaporise your base spirit with juniper and botanicals, or start from scratch with your own grain alcohol – so if we're talking serious gin, it needs to involve distillation.

I quite like a gin and tonic with a chicken jalfrezi, which may not be as strange as it sounds, since their botanical ingredients are virtually identical. Your little muslin bag of coriander, cumin, cinnamon bark and juniper berries would work as well in the cooking pot as it does in the distillery. So I wouldn't mind making a curry-based gin one day, without the heat but with the aromatics.

With the gin market as it is, it's probably being done as I speak, because there are now over three hundred gin distilleries in the country and about a thousand different gins on the market, so many that a lot of them probably won't survive no matter how good they are. I'm of a mind to buy for my collection the ones that I don't think will be around in five years' time.

When you're doing a distillation, different oils get released from the botanicals at different temperatures. Some might need a high temperature, others a lower. A dried orange peel, a juniper seed and a soft sweet cicely plant all break down at different temperatures, which is why you need the fixers and smoothers – the roots and gums – to keep everything in check and hold it all together.

When we were doing the recipes for the gins, we had the fixers at the base and the top notes from the botanicals – it was just like making a perfume. The different levels included the springtime tips of Douglas fir, sweet cicely, sweet flag – ingredients that brought a unique flavour to the whole brand. Gin needs juniper, of course, and we went to Scotland to get that. We were having to work through the seasons to gather the different components.

I used sweet flag, angelica root and orris root as fixers, to bind all the different flavours together and keep them in place. Orris root is often used as a base note for perfumes and is also one of the ingredients in Ras el hanout, a blend of spices and herbs that is widely used in Middle Eastern cuisine. It is a really classical fixer in gin, but sounded almost too exotic to be a British plant. Of course, when I thought about it, I realised orris is just another way of saying iris. You usually find the bearded variety in people's gardens, but I decided any sort of iris root would probably work quite well what with them all being related. You have to dry out the roots – which takes about five years if you do it old school. Fortunately a dehydrator can do the same job in a matter of days or hours, depending on how you slice the roots up. You then end up with something that you can pound down or cut into chunks. It has some really distinctive raspberry and violet notes, and smells a bit like those parma violet sweets. It's really potent; you don't need much at all. A few grams of each botanical is all it takes to make a distillation.

The essential oils used in perfume are incredibly expensive to make as you need so many plants to make such a small amount of oil. But it's actually quite simple to do at home – and unlike alcohol, distilling it is legal. The results can be spectacular. I keep thinking that I might try with elderflower, and have a go at making a perfume. I already have my own copper alembic still in the shed, so all I need to do is to get as much raw material as possible, stuff it in the still, top it with water, heat it up and wait for the steam to come out the other end. Oh, and try not to let it run out of water or it will blow up. Simple really.

Having sorted out the botanical base notes, I added the middle notes. First was myrrh gum, which is a smoother, binding the fixers together and soothing the root flavours. When I also added sweet cicely I had forgotten that the Latin name for it is *Myrrh Odorata*, so it could have been a bit of a myrrh overkill, but actually it worked out perfectly with the base and middle notes creating a bit of magic together.

I put together the other ingredients to create the taste by trial and error, but I'd built up a pretty good level of knowledge of how to blend botanicals, and what would work with what. While a lot of them have complementary compounds, almost all need to be treated in different ways to release the oils that produce their distinctive aromas and flavours. Some compounds are alcohol-soluble, some are water-soluble and some are oil-soluble, and it makes a difference. So you need to know at what point each one does what, in order for them to perform at their sparkling best; it's a bit like being an alchemist. It took me a whole season working with a guy who could do the distillation part. I brought what I thought would work and he distilled them and eventually I worked out what to bring to magic it all together.

The most important thing is how the gin tastes and how smooth it is, you don't want it be harsh. So, just like with the Christmas Tree Syrup the previous year, the first chef I took mine to was Craig Atchinson. He was more than delighted when I descended upon his kitchen with the latest samples and happy that he would be cycling home later on rather than driving. When it comes to taste you can't just rely upon your own opinion, you need the experts to tell you the truth and Craig gave me feedback that made a difference to how the gin was distilled. Having got Craig's approval, I sent a test batch to Michael Wignall. At the time Michael was head chef at Gidleigh Park in Devon, although these days you will find him at The Angel at Hetton near Skipton in North Yorkshire. Michael has a fantastic palate. Throughout his career he's achieved Michelin-star status for the restaurants he's headed up and like all top chefs, he wouldn't have got where he is today if he didn't have that palate. The other thing about Michael is that he's a Yorkshireman and so he says it how it is. He is brutally honest, which is what you need if you're asking someone to taste something for you. So getting his approval meant a lot.

Eventually I had created gins that I was happy with and we launched our range at Craig's restaurant at The Grand Hotel in York. My knowledge of botanicals had brought a unique flavour to the gin market and to my considerable pride and delight, one of the gins immediately won a gold medal at the Oscars of the spirits trade: the 2018 SIP Awards in San Francisco.

Later this year I will be releasing a brand new gin called The Yorkshire Forager, which will be made from an exciting selection of foraged ingredients that reflect all my years' experience in gathering botanicals from the British country-side and extracting from them the best authentic flavours. So watch this space . . .

CHAPTER TWELVE

ADVENTURES IN FORAGING

THE WISDOM THAT Grandad and others have shared with me and the experience I've built up over a lifetime means that I'm lucky enough to be able to make a living from the land, but my love of foraging runs much deeper than that. The word itself feels alien to me – it sounds like trendy opportunism and I'm as far from trendy as it's possible to be – but it's part of my family history and has made me the person I am today. It's not always easy, in fact it's incredibly hard work and even dangerous at times. I carry injuries and the battle scars to prove it. It can also sometimes be a monumental waste of time, when you can't find a plant where you expect it to be growing. But it's worth it.

It is how I earn my living, but it's not just a job. I love everything about what I do – and the fact that there aren't many folk who pick weeds for a living still puts a smile on my face. It's hard graft, and there are times when you're bent double cutting wild garlic, say, for hour after hour, until your back is screaming, it almost feels like you might as well be flat-packing furniture . . . except that you wouldn't be doing that out in the fresh air, revelling in the beauty of the wild, while getting a completely free chemical peel of your face.

People sometimes say to me, 'You call it foraging, but it's really just pillaging the countryside.' Well, it is and it isn't.

If you do it sensibly, it's more like gardening. If you prune a bush in your border it will come back twice as strongly next year, and it's the same with foraging. If you take anything, you thin it out, or only take part of it, and leave the rest so it can grow back. So it's all about being mindful. If something's really prolific you can take it by the roots, but in general you disturb the plant as little as possible. In order to take you must give back and over the years I've spoken to lots of people, including the local council, and made sure I had their support. These days I can be confident I have the permissions I need.

There are some edible varieties that I would never touch, not because they're rare, but because of the damage they do to our environment. Japanese knotweed was brought to Britain as an exotic plant in the mid-nineteenth century but is now one of our most damaging foreign invaders. It's almost impossible to eradicate, grows ferociously and forces its way through any barrier. It's illegal to pick it or move it as that risks it spreading and it'll devalue your house if you've got it in the garden. Yet I see foragers announcing online, 'We've been eating Japanese knotweed tips this week,' and I'm thinking, 'Well, what have you done with the rest of the plant?' because even if they've trimmed it down, the trimmings are liable to grow new plants. So whenever I see anyone sharing ideas of what you can do with it, I always post a picture underneath of the stuff growing straight through concrete, brick and tarmac, and say, 'This is why you shouldn't.'

You've also got to ensure that whatever you're planning to pick doesn't feature on the list of endangered and protected species. The Department for Environment, Food and Rural Affairs (DEFRA) produces a huge spreadsheet of them, and I'm one of those people who checks it religiously to see what you can and can't pick. There are a couple of hundred plants on the list, including rarities like hedgehog fungus – so called because there are 'spines' like a hedgehog's on the underside of the cap – and devil's boletus, with its blood-red pores, bulbous stalk and absolutely terrible smell.

While a lot of people I meet like the idea of eating wild food, they often complain, 'But it's really expensive, isn't it? Why does it cost so much?'

Well, you'd know why if you were risking life and limb to get the stuff. They're right that wild food is not cheap, but with good reason – foraging is hugely intensive and time-consuming. You don't have to be massively fit – I'm the built-for-comfort-not-for-speed variety of Yorkshire woman and you won't find me running marathons or pumping iron at my local gym to prepare for the foraging season – but it's hard physical work, and often in very difficult conditions.

We don't have plants growing in polytunnels or on an allotment, they're all out there in the wild and the things we have to go through to get some of them would make your eyes water. At the peak of the growing season we can be out from dawn to dusk and beyond. We have falls, we get injured, we're out in wind, rain, hail, sleet and even snow, we get bitten and stung constantly, and our hands and arms are always covered in cuts, bruises and scratches. I'm an expert in untangling myself from barbed wire, razor thorns, spiky bushes and the odd illegal animal trap.

I don't do hands-on courses for people who want to learn about foraging, though many others do, because we are finding and picking things in fairly large quantities for restaurants and we tend to have to focus on one thing, in one place, at a time. For example, if we're gathering Sitka spruce tips – the tiny little buds on Christmas trees that are really fresh and lemony when they first appear – it'll take us ages to get a reasonable number. It isn't a thrill a minute, so no one is going to want to come with us while we do it. Just standing in front of the same tree for hours on end doesn't make for the most riveting spectator sport – but doing it can be seriously good for the soul.

To do the job, you obviously need a considerable level of knowledge about what you're looking for, and you need good eyesight. The other essential is an obvious one: to know where to find things. Part of my job is knowing how to track down rare plants, and especially mushrooms, and there is a methodology of how to do it. It's not just about the environmental factors: the right habitat, temperature, rainfall, altitude, orientation, soil pH, and so on. If you're trying to track a plant down in a wood, new woodland isn't that productive, so ideally choose an ancient one. I'm lucky that there are a lot of those around us, including bits of Sherwood Forest that have been virtually undisturbed for a thousand years or more.

Having found your wood, the first places to look are always the paths. Animals and humans all use the same paths, so we're always brushing against plants and carrying their seeds up and down them with us. A walk in the woods usually results in us coming back covered in burrs and seeds with hooks in. I have to hoover every day and it's a bit like vacuuming a seed store. I'm surprised my carpets aren't like fields. Mushrooms are really sociable so they like to be on the side of paths too, and they always have friends; if you find one, you'll always find another one nearby. Some grow in troops and some are lone wolves, but even the lone wolves will have others growing near at hand.

Sometimes when you're searching for plants, other ones might give you a clue as to their whereabouts. Holly trees are pretty random; they can grow in quite warm conditions and you'll often find them deep in the woods, but they can also thrive up on the moors, in sub-arctic conditions where very few other trees can survive, and you will always find something interesting growing nearby. I think they may be a sign of particularly fertile soil and my mum certainly believes that, because whenever we are out foraging together, she always makes a beeline for a holly tree as a starting point.

As you become more experienced at foraging, it almost becomes more of a science than an art. For example, leaves have patterns that are always uniform and replicated through the entire plant, so any leaf taken at random will be exactly like its sister leaf. Look at plants with that in mind and you will be able to recognise those patterns and formations and distinguish between different plants at some distance from them.

You'll have your regular spots, but when you stumble on a new wild plant or a new source of the ones you already gather, you need to make damn sure you can find it again, twelve months later, when the next crop is ready. Although I remember where I found most things, I keep a record of the precise locations of all the plants I've been foraging, even down to the longitude, latitude and GPS details, for fear I'll forget by the time the next season comes around. I also make a note of what conditions they need to thrive, what time of year they come out, how long they last, and so on, from one week to the next, because nature stops for no one.

Luckily I'm good at retracing my steps, or at least I am when I'm out in the wild. When I'm on the roads or in a city my sense of direction is appalling – I need a sat nav to drive a mile down a straight road – but put me in a wood, no matter how dense, and I can find my way out. I'll tell myself, 'You need to turn right at that stump, left at the copse of birches, straight on past the burnt yew, take a left by the pignuts and then it's two hundred yards on your right, by the tree with the King Alfred cakes on it.' That would probably be meaningless to anyone else but it makes perfect sense to me and is usually enough to stop me getting lost.

I instinctively seem to know which route to take, even if it's a wood I've never been in before. I just recognise the way the trees lie and can find my way back to the car, but all the time you have to be keenly aware of where you're walking because it is easy to get disorientated; the old saying 'You can't see the wood for the trees', is often literally true.

If I'm in unfamiliar surroundings I'll sometimes take a compass, because even in local woods, you can be led astray. There's one near where I live, what I call a scrub wood with no giant oaks or anything striking that you can use as a marker. In spring there are St George's mushrooms as far as the eye can see and if you're not careful, you can be so excited, and so focused on picking them, that you can lose track of where you are. So a bearing can be quite handy, though leaving my bright pink rucksack dangling from a high branch I've passed on the way in doesn't half help too.

We live in such a noisy society that it's rare not to have the sound of road traffic or the flight path from your local airport to guide you, though being able to tell your north from your south, east and west will always come in useful.

Not many people can do that without a compass, but there are plenty of clues if you know where to look and what you're looking for, from the position of the sun at different times of day, or the stars at night, to the fact that moss grows thickest on the north side of trees, snow will stay longest in north-facing slopes and gullies, and so on.

I've always liked to head out after dinner – that's proper Yorkshire dinner, which some people call lunch. To this day that still confuses my friends around the country and I have to explain: 'Up our end, we have breakfast, dinner, tea about six p.m., and supper around nine, which makes it sound like us northerners do nothing but eat. Down south, it's breakfast, lunch and dinner. In Hampshire and the south-west, it's breakfast, lunch and supper. And if you're in the Navy, every meal is just called "scran" and every drink is "a wet".'

There is a very good book called *Jackspeak* which is all about Royal Navy slang. If Ryan ever decides to follow in my footsteps and join the Navy I'll buy it for him. He wouldn't appreciate it until he went in, and even after a year trying to work out this new and strange language, he'd probably know sweet F.A. about anything. Sweet F.A. started out as Navy speak for much-disliked meat stew. Fanny Adams was a young girl who had been murdered in 1867; her body was cut up and it was alleged that her remains were found in a Deptford victualling yard around the same time that tinned mutton was introduced into the Royal Navy diet. Over the years sweet F.A. changed to mean anything that wasn't very good, and then eventually nothing at all. I wish I'd had the book when I joined up.

There was a time when I spent so much time in the woods that it was second nature to be out and about in them and it didn't bother me if I didn't get back home until dusk, whether it was a freezing cold Boxing Day or high summer, when dusk can be as late as ten p.m. Over the years a few things made me change my mind, and these days I usually send Chris out to do the picking at that time of day because I've had too many encounters with random strange men, some of them downright alarming.

I was coming back through the woods one evening, carrying a bag of wild raspberries I'd picked, and as usual, Fred was a couple of paces in front of me. Suddenly a man jumped out of the bushes, wearing nothing but a pair of bright orange Speedos and with his hands in the air in a sort of 'Ta-daaa!' gesture.

He was only about five foot six inches tall, pretty skinny, and clearly not armed with anything, so I wasn't too per-turbed and kept on walking towards him. As I did so, I noticed that he was surrounded by a very dense and hungry midge cloud, so without missing a step, I said in my best Yorkshire accent, 'You're going to get bitten to death by those midges if you go around wi nowt on like that, love.' Both he and his Speedos looked rather deflated by that and he disappeared back into the bushes as I walked straight past him. Fred didn't bat an eyelid. There was clearly no threat as, despite being as bright as a 20-watt bulb, Fred has excellent instincts.

Another night I was out late and making my way back to the car through a wood with a very slippery paved path, so I was more intent on watching my feet than seeing what was going on around me. I didn't have Fred with me this time and was dressed in boots, odd socks and a gilet, with my hair in a scruffy ponytail, no doubt with a few bits of twig in it, so

I wasn't at my most appealing. A man hurried up beside me, but then slowed his pace to match mine. That immediately put me on edge, but I kept looking at my feet rather than at him.

'I've just finished work,' he said.

Still without looking up, I said, 'That must be nice for you.'

'I live on my own.'

'Do you now?' I said, still looking at my feet but walking a bit quicker.

'Where are you going?'

'My husband's waiting for me at the car park.' He wasn't, but I felt worried enough to say it.

'Oh right,' he said. 'I'll walk you there.'

'It's all right, no need,' I said, but he just kept on walking alongside me.

When we reached the car park, as he was looking around, no doubt trying to spot if my husband really was there, I walked straight past my car, then did a very quick U-turn, dived into it and locked the doors. He just stood there looking at me as I drove off. You just never know. Well, sometimes you do, and then you need to be quick witted.

From then on, I always took Fred with me if I was in the woods on my own towards dusk. One evening, Fred, who is a giant of a labrador, weighing about forty-five kilos, but normally a very placid dog, kept turning round to stare back down the path behind us, and his hackles were up. A few moments later, a man appeared, walking very quickly towards me, but as soon as I made eye contact with him and he saw that I had Fred with me, he stopped and veered off sharply, disappearing into the bushes. I've no doubt that Fred had saved me from being mugged, or worse. His instincts had kicked in and he knew this was a man with bad intentions. Fred has his uses, bless him.

The cumulative effect of those unsettling experiences was to make me much more cautious about going into the woods in the later afternoon and evening, and if we need to gather plants or mushrooms at that time of day now, I either take Chris with me or send him on his own, and always with Fred for company.

You also never know *what* you're going to find in the woods, and it's not always wonderful wild food. I was out one March, picking the first few shoots of wild garlic. I needed to gather quite a bit so I took Fred with me for the walk, he likes nothing more than a trundle. Like all dogs it's the smells he loves most and on this occasion Fred's nose got the better of him, as it often does – nothing is more exciting for Fred than smelling a dead badger three fields away. On this particular day his chocolate nose went up in the air like the Bisto kid and he was off, but not too far and on the path. After a bit of a rummage in the undergrowth he came out carrying something bright blue and looking very pleased with himself. I got a bit closer to him and with some disgust I realised that the prize catch he was carrying like a stick was actually a sex toy. He wouldn't come when I called, and as I approached him I saw the whites of his eyes and his ears flatten – his new-found sex-toy stick was even more precious now it was clear that I wanted it too.

Fred kept a few paces ahead of me and carried it proudly for three miles, when we came upon a local ramblers club. The ramblers smiled as he approached them, but their looks quickly turned to disgust and they parted like the Red Sea to let Fred and his sex toy through. I followed after him doing a walk of shame through the silent ramblers, quietly protesting that it wasn't mine and he'd just found it. Sometimes you just don't know what the day is going to bring.

The idea of foraging may conjure up an image of someone prancing around the hedgerows, snippers in hand, collecting a little bit of this and a little bit of that, and placing it reverently in the basket or trug on their arm, but it's nothing like that with me. Not only am I more likely to be accompanied by a dog carrying a sex toy, I certainly won't be carrying a trug as I don't use them. I buy giant green carrier bags and the great big green bin liners you can get from Home Bargains, about ten for a quid. I use plastic bags because they keep what I'm picking fresh, but I use them repeatedly and recycle them when they're knackered. I've got a carabiner hanging off my belt to hold all the things I need to carry with me, because when I set out, I never quite know what I'm going to find. I have foraging nets for bilberrying, a sixty-six-litre rucksack for sea herbs, and a bright pink tiger-print shopping trolley that I use for puffballing. By the time I emerge from the undergrowth I look like a fully laden donkey.

I use all sorts of sharp implements to cut the plants I gather. I've a steak knife for doing wild garlic, a knife with a ten-inch curved blade, a bit like a machete, for hacking away at mugwort and sea aster (it saves me bending and wrecking my back), a curved mushroom knife so I don't ruin the mycelia, and a host of others. I've got knives in the glove compartment, the side compartment, and the central and overhead compartments of my car. They all have a specific purpose, but I really do have a lot of them and, although I assume that, like a chef, I wouldn't be arrested on the spot if stopped by the police, I've never actually put that theory to the test.

Unlike the Speedo-wearer, I never want to alarm people who see or encounter me when I'm out foraging and I reckon the chances of that would be an awful lot worse if I was dressed like a tramp on a bad hair day, so I always apply full makeup, put my hair up and dress in good-enough clothes before I venture outdoors. I'm quite partial to a sequinned top and all, in fact anything sparkly or metallic. I hope it makes me look nice, if that's possible. A lot of the time we have to lurk about in the undergrowth and it's a case of deciding to go full SAS combat style, or just being up front about it – and for me that's the best approach. I'm the Yorkshire Forager, not a Ninja warrior. So it's shiny stuff and pink lipstick to go with the machete, and I even got Chris to give up his striped nylon jogging bottoms, because I thought they made him look like some kind of Eastern European gangster.

SOMETIMES THE DANGER we put ourselves in is of our own making – not deliberately of course. When we are up in Scotland gathering juniper berries, my mum and I will sometimes go around a couple of whisky distilleries – as you do, what with her and Chris being rather partial to the stuff. She always ends up bringing quite a few bottles back and putting them in her booze cabinet, never to see the light of day again.

My brother has already said he wants first dibs if anything ever happens to her – that's not first dibs on the ornaments or jewellery, it's first dibs on the booze cabinet. I have to say my mum is amused by this. We aren't afraid of death in our family and we often discuss how we can do it on the cheap. Grandad wants me to pick him up in my van and take him to the crem in a cardboard coffin and Mum fancies a wicker one, but I think Nana wants something a bit more dignified. I think we can talk like this because we have a lot of love for each other, and lots of memories, and we know that both will last forever.

Once Mum and I went to a distillery that had its own cooperage, where they made the oak barrels used to age whisky. We watched the cooper assembling the staves and then encircling them with red-hot steel hoops. It was a grand show, with clouds of oak smoke filling the air as the hot metal scorched the wood and then billowing clouds of steam as his apprentice tossed a bucket of water over them, causing the metal to contract and bind the staves together so tightly that not a drop of liquid could escape from the finished barrel.

While we were watching him, ever alert for a bargain, Mum noticed that there was a pile of old barrel staves piled up in a corner and said, 'Excuse me, what do you do with those?'

'We've got houses on the site,' he said. 'And the people use them for firewood.'

'So could we?'

'I'm afraid not. We have to keep everything in house.'

Not to be defeated, I sought out a cooperage in the Highlands that supplied barrels to a lot of the really big distillers, and when I got in touch with them I discovered that every so often they disposed of barrels that were so old and saturated with whisky that they had begun to leak. This time, when I said, 'And could we get some?' the answer was, 'Sure, how many do you want?'

'As many as I can fit in the car.'

'OK,' he said. 'We're emptying some very special fifty-year-old barrels on Friday, so if you come round first thing on Saturday morning – about six o'clock – I'll have some sorted for you.'

So we put down the seats in the back of the car and went round on the Saturday and loaded up with staves, piling them up until they were the height of the seats. If we'd stuffed any more in and I'd had to brake in a hurry, the whole lot would have come out through the windscreen, taking us with them. We packed them as well as we could, using sackfuls of juniper branches to wedge them in.

We parked outside our hotel overnight and when we got into the cab the next morning, the whisky fumes were so strong they nearly knocked us sideways. I had to drive all the way back to Yorkshire with the windows open and my head hanging out, and if I'd been stopped and breathalysed, it could have been an instant ban. I felt alright, but I had no idea how the fumes might be affecting me.

When we got home with the whisky chips I bought one of those huge dumpy bags from our local builder's merchant and asked a friendly tree surgeon to put the whole lot through his wood-chipper. He did it at the end of our street and the fumes were so powerful that his eyes were streaming and the whole place smelt like the Highland on Hogmanay. That wasn't a bad moment for any of us.

I now had at least a ton of fifty-year-old whisky chips that would be great for barbecues, but I was never going to use them all, so I drove round to see my chefs with the surplus and invited them to 'just come and sit in my car and have a good sniff'.

How could they resist?

They all took some of the whisky chips, and some of them told me later that they'd 'brined' them – put them in water to leach out all the top-notch single malt of an age that would have made it far too expensive for mere mortals to purchase – and then soaked salmon and seafood in it; pure genius, but that's chefs for you.

Apart from the risk of being breathalysed as we were driving round with the whisky-soaked wood, there was another hazard, but not one that occurred to me, Chris or anybody else who had been inhaling that intoxicating aroma until several years later. I was round at Nana and Grandad's house one afternoon and was regaling my Aunty Marilyn with the story of the whisky chips. When I'd finished, she said, 'Good job you don't smoke, Alysia.' I looked at her a bit confused but then it hit me like a sledgehammer. We'd been driving round the north of England in a bomb on wheels. Literally. It gave me the shivers when I thought of what could have happened. It was only by the grace of God, so to speak, that we hadn't blown ourselves up.

I thought often about what Aunty Marilyn had said for weeks afterwards. On one hand I couldn't believe that I hadn't thought about the flammable risk, and on the other I could understand that at the time I was more concerned that I might be getting drunk from the fumes. In hindsight, I remember the look on the bloke's face when we turned up in our Citroen Berlingo, and not an enclosed Luton van like he had told me other people used. He might have said something at the time! Oh well, you live and learn. Would I do it again? Probably . . . but in a Luton next time.

When I think of my family, we have always been the sort to give things a go. It might not work out but they say God loves a trier and my mum has always said I've been very trying, bless her. Nevertheless she has always stood by me and been willing to go along with my plans. I think she likes a challenge and especially if it's fun; she was as excited about finding the whisky staves in the Highlands or the tobacco bales in Cornwall as I was. Chris on the other hand, has a bit of a moan but then he gets on board and sticks by me. And even though both of them are rolling their eyes at me, they have learnt by now that usually I do have a grand master plan and it does generally come off.

CHAPTER THIRTEEN

CHANGING NATURE

WHEN I WAS YOUNG, we used to swim in the River Ryburn, which flowed through our village. It was minging, quite frankly, but we still did it. No one talked about conservation back then, and in the eyes of most industrialists of that era, rivers were just for getting rid of all the crap from their factories. I can remember walking along the bank one day in the 1970s and all the trout were floating upside down on the surface. Probably a dye works or a tannery or some-such had discharged something deeply unpleasant into our paddling pool, or some ghastly pollutants had been leached from the soil by heavy rain.

We think we're much more environmentally conscious now, and in some ways we are, but some modern-day farmers are just as bad as those old industrialists with the slurry and the silage run-off that they allow to invade our streams and waterways, and the pesticides they spray on their crops. I see the effects of this short-sightedness each and every day, because when you're out in the countryside, in the fields and woods, by the rivers or on the moors, you have a finger on the pulse of the natural world, and you can see for yourself how things are steadily disintegrating. That's not to say that farmers don't do a lot of great work in conservationism but it's what slips through the net that can cause the most damage to sensitive eco systems.

Intensive farms still use pesticides like there's no tomorrow, and despite the agrochemical industry's claims they are safe, I am convinced they are an absolute menace. I see the lack of bees, the lack of insects and the lack of plant diversity, and it frightens me. We need all these things. I hate clouds of bugs around me, midges and sandflies biting me, wasps, bees and hornets stinging me, wood flies driving me mad by hovering just behind my head and buzzing past my ear every couple of seconds – but I know we have to put up with them because they are part of the fundamental essence of our lives.

Every insect has its role to play, even if it's not always obvious. Take the tansy beetle, for instance. It's native only to the area around York and lives on the tansy plant alone. It can fly but doesn't, so it walks a distance of up to two hundred yards to find another plant to feed on, and if it can't find one, it dies. How many other insects have we lost that we didn't even know existed? To me the tansy beetle is the same as the giant panda: it's beautiful, it lives in limited locations and only has one food source. But not many people are falling over themselves to save it. I believe all things should be saved; Nature put them there in the first place and they are important to the planet, if not to us.

One glance at a hedgerow will tell you that all sorts of plants are thriving despite the multitude of insects in residence, but there won't be a single bug in a field of wheat because the crop is regularly sprayed with pesticides and selective herbicides to keep them away and suppress other plants. Other than the all-powerful mono-crop, the entire area becomes barren of life. Meanwhile the toxins in the pesticides are busy working their way up the food chain. Along with climate change, insect diversity is absolutely critical, and yet, from what I see with my own eyes, it has deteriorated rapidly and continues to do so.

People have become aware of the importance of bees to our ecosystem and it is good to see initiatives in place to protect them. When I see those campaigns, I always think of the bees I save when they are exhausted. I keep honey on my window ledges at home; it's like rocket fuel for bees. I've seen a bee suck up three times its body weight and go from lethargic to superhero in the space of half an hour. Off they fly, leaving behind a small, crusty ball of sugar after they have sucked all the goodness out. It's absolutely fascinating to see them recover so well.

It's not just the insects that are under threat. Some species of animal, bird, fish, invertebrate and plant are in what may be terminal decline, while a handful of others, such as poisonous dog mercury and ferns, are taking over every-where – and hastening the process by which the richness and diversity of the woodlands and wild places I knew when I was young are changing rapidly.

The hares I used to watch boxing when I was a kid are much rarer now, a decline that I suspect has been fuelled as much by the loss of habitat caused by changes in traditional farming methods as by pesticide use. Hay meadows were once always left undisturbed until the crop was cut, usually no earlier than late June or into July, by which time hares and their friends and neighbours, and ground-nesting birds, had all finished rearing their young. Nowadays, farmers who have switched to silage as winter feed for their cattle and sheep can get two or even three crops a year, beginning in early to mid-May. As a result, young hares and the offspring of curlews, lapwings, skylarks and partridges fall victim to mowing machines and mechanical balers.

Even worse, except for a few areas like the uplands of the Yorkshire Moors and Dales – where, encouraged and sometimes subsidised by the National Parks, many farmers continue to grow and cut hay in the same way that their Norse ancestors did – the majority of those herb- and wildflower-rich meadows have disappeared. They had been growing undisturbed for a thousand years, but most have now been replaced by rye grass, which gives farmers heavier crops but has none of the richness and diversity of traditional meadowland and supports far fewer bees, butterflies and other useful insects.

The loss of gardens and the increase of house building means water has nowhere to go, so it congregates at the bottom of valleys, where rivers and canals burst their banks on a much more regular basis. I experienced the impact of this first hand, on Boxing Day in 2015 when we woke to find the steps round the back of my mum's were a gushing waterfall. We needed to get back to Doncaster, but the river in Sowerby Bridge had risen and access was completely blocked for the first time in a hundred years. Chris and I set off for Ripponden but came across another flood. I racked my brains for a route, and decided to go to the tops of the moors, but we came to a small bridge going over a culvert that was also flooded. I stopped the car to assess, and while doing so another car came the other way and managed to get through. So with a deep breath I went through. We managed to get through and back to the dual carriageway and home. But I had never been so scared in my life. We were lucky to make it home – the bridge we had crossed had collapsed just minutes afterwards.

People are encroaching ever more on the land, often with disastrous consequences. The dry-stone walls around my home get regularly plundered for building material or rockeries, fly-tippers – some of whom just bin their beer cans and pizza boxes in the nearest ditch or empty their car ashtrays at the side of the road, while others have graduated to fridges, soiled mattresses and dilapidated three-piece suites – are everywhere, rare plants get dug up or trampled underfoot and wildlife is disturbed, driven off or even wiped out.

As kids we used to collect newts and frogspawn, which probably wasn't very ecologically sound of us and is against the law now, but there were hundreds of newts and gallons of frogspawn back then. Now, when I go to Norland to see Mum and Nana and Grandad, although I still haunt the same places and wander through the same woods and moors as I did then, I can't remember the last time I saw even a hint of them; all the swampy areas where they used to live seem to have dried up.

The loss of such creatures is perhaps another consequence of climate change, but it's also a reflection of the impact of an ever-growing population. New houses are built, more people take walks through the woods, more paths are created and even the supposed defenders of the environment can do more harm than good. If you complain, they'll have lots of people with degrees in environmental studies to argue that what they're doing is right, and they wouldn't think of listening to people like me, who actually know these areas more intimately than they could ever imagine. My memories of the land go back forty-five years and my family's memory is passed on to me, so I know what was there before me and I know what is there now.

A Wildlife Trust might be left a piece of wild woodland in someone's will or take over a local beauty spot. So they'll make a nice accessible tarmac path straight through the middle of it, and they'll tidy up and gravel all the meandering little trails which used to lead to our favourite hideaways then they'll probably clear out a wonderfully diverse tangle of plants in order to make a grassed picnic area, complete with benches and tables, or a children's playground. They'll clear another acre or so for a car park and before you know it, there'll be a tea room and you'll be charged four or five quid just to be there. If you challenge them, they'll say, 'It's because we're looking after the environment.'

Which makes me want to reply, 'Well, before you came along, the environment had been managing perfectly well by itself for the last thirty thousand years.'

I've seen them do it. There used to be a lovely stretch of aniseed toadstools near where I live – despite being called toadstools, they are edible, but not at all common. Then one time I found they'd cleared the area and put in picnic tables and a climbing frame and slide. Children need to play, of course, but I would rather we were catering for them in ways that encouraged their imagination to soar, and weren't so damaging to the local flora and fauna. It absolutely drives me nuts, because we need to surround ourselves with wild nature in all its richness and diversity if we're going to survive ourselves. Humans cannot stop interfering, thinking they know best all the time instead of letting things just be natural and Nature take its course.

We tend to take our prosperous, secure way of life for granted, with no end of central heating, cars, computers and electronic gadgets, but we may only be one step away from catastrophe. If some terrible crop failure or environmental disaster was to strike us, most people would be running to the shops to buy bread, bottled water and tinned food. I wouldn't. I'd simply liberate some salt, vinegar and oil from my store-cupboard, because I can find everything else I need to survive in the great outdoors.

There are things that are edible, like apples off a tree, that everyone knows about. There are things you can eat, but need a bit of processing, and most people might not be aware of them. There are other things that aren't particularly palatable, but you could eat them if you had to in an emergency; I'm thinking of the ash keys, acorns and pine needles that Grandad foraged in the Polish forest during the war, just to stay alive. Even fewer probably know about them. And then of course, as we have seen, there are the things that you need to avoid because they will make you ill or even kill you. Some are familiar – most people probably know not to eat fly agaric or foxgloves – but a lot are not. So you need knowledge to help you identify what to eat and when.

The rise of forest schools for infants and young children is a great idea, starting them off young and getting them used to nature. They're already heavily over-subscribed and I think the government should be funding more of them, so more kids from all backgrounds can learn more about the environment we all share. I suspect that if you took the vast majority of people into a forest, they wouldn't be able to tell a conker from a sweet chestnut any more than my teacher could back in the day, nor know the names of the trees or what the bark looks like – why one is a pine and another an oak – nor what parts of which are safe to consume. In fact, there aren't many trees that are poisonous, holly, spindle and yew being the most obvious examples, but there are lots that you could eat.

When I was jumping through the hoops to sell my Christmas Tree Syrup, I did a course to get a Level 2 Food Safety and Hygiene for Catering qualification. It was a requirement if you wanted to get Safe and Local Supplier Approval (SALSA) certification, a scheme operated by the Institute of Food Science and Technology for small producers looking to supply food retailers. In her introduction, the course leader told us that in the old days, people just got a bar of soap and did the necessary with their food preparation areas and the utensils and equipment they used. It wasn't anti-bacterial, it was just a good wash, but everyone was happy with that.

Not any more.

There are still rustic-style restaurant kitchens with a one-star hygiene rating, but there are others with five-star hygiene ratings which are gleaming and immaculate, with digital controls and sensors, and more all-singing, all-dancing gizmos than you can imagine. There's not a microbe to be seen, but that isn't an entirely good thing. Global corporations are constantly trying to boost their profits by selling us yet more stuff that we really don't need, and they've managed to make us believe that any bacteria are a problem. That doesn't actually do us any good, because we need microbes and bugs to build our immunity levels, and we don't need anything fancier than a bit of good old-fashioned soap and water to keep us clean.

Our bodies have spent thousands of years acclimatising to the planet we call home, but we now surround ourselves with plastics, carcinogens and pesticides, and use biological washing powders, hand sanitisers, anti-bacterial cleaners and the rest, and then wonder why we've all got allergies and are ill every five minutes. I get far fewer coughs, colds and stomach bugs than other people, and I'm sure it's because I spend a lot of time scratting around in the dirt, in the fresh air.

There are so many good bacteria out there, and you can't fight bad bacteria if you don't have good ones in your corner. So in trying to make ourselves safe, we might actually be putting ourselves more at risk, because we're reducing our natural immunities and our ability to cope with what's out there. I'm not one of those 'The End of the World Is Nigh' doom-mongers, but you can see the warning signs, and if humanity keeps ignoring them, for sure we're going to pay the price in the end. Some virus or disease will evolve or mutate, something even more lethal than Ebola or Aids and it will be fast and fearsome.

Everything on the medical roster ultimately derives from the natural world, and there are countless researchers closely examining the properties of rare plants from around the world, from the depths of the Amazon rainforest to the remotest parts of the Siberian tundra. Yet we should be devoting as much and more funding and research effort into looking at fungi and lichen. Even at current levels, two thousand new species of fungi are being discovered every year, but we're barely scratching the surface; it's estimated that over ninety per cent of all fungi still remain undiscovered.

The possibilities are limitless. Scientists in Pakistan have grown a fungus – *Aspergillus tubingensis* – that can ingest plastic, turning a terrible pollutant back into a living, biodegradable organism. There are also bacteria that can degrade plastics, and even two species of waxworms, which, as the name suggests, normally consume beeswax, but which can degrade plastic as well.

So the plastics that are such a problem for us, that originate from some of the earth's natural resources – crude oil, coal, natural gas, minerals and plants – could be converted back into natural, biodegradable substances just by using the right enzymes, bacteria, fungi, or invertebrates. That's where all the research effort should be targeted, because it doesn't matter how many recycling bins, litter-pickers or giant ocean scoopers you've got, you are never going to get rid of all those plastic containers and water bottles which, by the way, should already have been banned. Nobody had a plastic water bottle when I was a kid; we all managed to get along without them perfectly well, and I'm sure we could again.

As well as plastic-eating fungi, an anti-tumour drug has been developed from a chemical found in giant puffballs, and biological detergents were developed from fungi, but there are so many thousands of mushrooms, containing so many hundreds of thousands of compounds, that there must be many more little miracles waiting to be discovered. We just have to do the research, but we need to do it soon, because the natural world we all share is being imperilled at an ever-accelerating rate, as not just any forager, but anyone who looks at it with an open mind, can testify.

CHAPTER FOURTEEN

YORKSHIRE FORAGERS

SHEPHERDS IN THE Yorkshire Dales talk about their sheep being 'heafed' to the particular area of moorland where they will spend their lives and from which they will not willingly stray. Indeed, so strong is the pull of the heaf that when upland farms change hands, the flock is always included in the sale. I feel a similarly powerful attachment to the Yorkshire fields, woods, moors, hills and valleys that I have known all my life, the places where my grandad began to teach me the foraging knowledge and skills that I've continued to draw on to this day as I work alongside my mum.

Grandad came through some of the darkest years in our history, and survived in a very unusual way. In order to do so, he had to make tough choices every single minute of every single day. Throughout the rest of his life he has lived very simply; he must have wanted a quiet life after what he'd been through, and he'd certainly earned one. I was round at his house a few years back when an official-looking letter arrived for him. It was from the Polish government, offering to send him the medals he had won fighting for the Resistance. I think Aunty Marilyn had looked into it at the time and we were all excited about this belated acknowledgement of the role he had played against the Nazis, but Grandad said nothing for quite a while, just staring at the words in his native language, which now belonged in another life, long ago.

At last he stirred, and looked up. 'I don't think so,' he said. 'I'd rather let the past lie.' He's never really spoken to us or, as far as I know, to anyone else other than Nana about what he went through during the war, and he must have compartmentalised a lot of things just to be able to get on with his life, because there was no psychological counselling or treatment for post-traumatic stress disorder or anything like that back then. As people used to say, whatever hardships and traumas life threw at them, 'There's no point in moping. You just have to get on with it.' And that's what Grandad did.

The only visible signs of what my grandad has endured are his obsession with food and his need to make sure that there is always enough to eat in the house. He's been like that all his life and still is now – he's always checking they haven't run out of bread or buns. Nana does get fed up with it sometimes, especially when he starts cooking tea at about eleven in the morning, and then has to reheat it at five o'clock when Robert – who still lives in Norland with them – comes in for tea.

Back in the days when he went to the market in Sowerby Bridge every week, we all had to sit in the living room while he showed us the food he'd just bought. He'd stand there with a big smile on his face and say, 'Just look at the size of this cauliflower!' Once we'd made all the right noises about it, he'd disappear into the kitchen and come out again with a huge cabbage. On and on it went until we had nodded approvingly at every carrot, courgette and cucumber. It was never-ending, and we wound up eating an awful lot of vege-table stew. Even now, if he's up to it, Grandad's favourite outing is to get my mum to take him down to Lidl so he can admire all the food and buy things he doesn't really need, but can bring home and show us.

Like a lot of other people who went through the war and experienced years of rationing and real hardship, Nana and Grandad have never wasted a thing. Long before recycling had even been invented, they had drawers full of carefully folded brown paper bags, bits of cardboard and neatly tied lengths of string, and tobacco tins full of paperclips and rubber bands.

Even when rationing was at last over in the mid-1950s, Britain wasn't exactly a land flowing with milk and honey, and the range and variety of foods we take for granted now simply didn't exist back then. Nothing was air-freighted – and no one could have afforded it even if it had been. Even when I was growing up in the 1970s I can remember the greengrocers never had much more than potatoes, turnips, carrots, onions, cabbages and sprouts on their shelves in winter.

Nana and Grandad are now ninety-two and ninety-five. They'd been married for seventy-one years on Christmas Eve 2019 and still live in the house where they brought up their children, halfway up the valley, on the edge of the woods that lead up to Norland Moor. They still have their son Robert living with them, and of course Mum is now next-door-but-one, in what was Aunty Dora's house, at the other end of the row of cottages. Aunty Marilyn comes over every two weeks to give the house a good bottoming clean, which they really appreciate, and they love to see her. They are well looked after.

Everybody who knows them loves Nana and Grandad and when they were younger they would do anything for any-body. When Aunty Dora lost her husband, she became quite infirm and couldn't look after herself very well, so Nana went in three times a day to cook her meals, make sure she was washed and dressed properly, and generally look after her. For a while, two elderly sisters lived in the middle house in the row and, just as he did for Dora, Grandad did any heavy jobs in their gardens and swept and washed their steps. He even built a new set of steps to make it easier for everyone to get up to the communal garden that the three households shared. Grandad isn't a man of many words, but he enjoys a good laugh, an all-action film and a bit of flat-pack furniture assembly or DIY. In particular he does love to build something with cement, and it's a trait he's passed down to my mum. They're always getting the cement out and building things in their gardens. The steps were a bit dodgy, though, and would keep falling down, but that just gave Grandad the excuse to make more cement and rebuild them. And of course he always shared the fruits of his foraging in the woods with them. He is an absolutely lovely man.

Grandad has always been shy and reserved, reluctant to talk much about his incredible story. He's more of a quiet observer of what's going on around him, often with a faint smile playing around the corners of his mouth; he's always been a very good listener. Even in his nineties, his mind is still razor sharp, though he's now lost his hearing, which obviously makes communication more of a challenge. I have to scribble notes to him and hold them up for him to read and then he'll reply in broad Yorkshire, but still with more than a hint of his original Polish accent in the mix.

Sadly, he's now very frail. When I went to see him recently and asked him how he was, he said, 'I'm doing badly. Do you know what that means?'

'Not exactly, Grandad,' I said.

'It's what they say in Barnsley. I'm doing badly now.' And given the Yorkshire love of understatement, that meant he really was ill.

Although Nana, Mum, Aunty Marilyn and I have pieced together some of his story, it seems that he will take the full details of what happened to the grave. But he has never forgotten how much he owes, including life itself, to Nature's richness. He knows only too well how harsh the world can be. It pleased him when we took notice as children to what he was telling us about the plants and the trees, and if we remembered it at a later date it pleased him a lot. We all have the knowledge to survive if everything went a bit pear shaped. I often get people saying to me, 'I'm coming to live at your house if owt goes wrong in the world.' Well that's good as long as they like wild garlic. We never used to appreciate the knowledge, but we do now. So he has done a good job with us all.

As I walk today through those woods and moors he loved so well, I often think about how his past has become my present and my future. When I walk on the ancient stone slabs, laid long ago when these now overgrown tracks were used by packhorse trains and travelling pedlars, I can still hear the echoes of Grandad's footsteps and the words he said to me all those years ago.

Last time I was up there I walked past what I still think of as Mr Sekulka's farm. Mr Sekulka is now long dead, but his Polish origins and the circumstances that brought him to Britain were similar to Grandad's, and when I was young we would always pause there. While they chatted in Polish, I would admire his horses and his Western riding saddle, just like the ones I'd seen in those John Wayne movies. When we moved on, we would catch sight of stoats disappearing into the dry-stone walls as we climbed the steep, cobbled tracks, and near the moor edge, the wind would often make us catch our breath as we came out into the open.

Whether you're in Yorkshire or any other rural landscape, you soon realise that Nature is never still for an instant. Tomorrow always brings something new and I live for those changing seasons, forever refuelling my senses with joy and wonder. I know where and when to find wild raspberries, all types of native truffles, wild sweet black cherries, wild plums, acres of St George's mushrooms, puffballs, morels, ceps and chicken of the woods – and you have to know your spots for those because they're biennial and only appear in alternate years.

In my local woods, I know where the flowers and plants come out first and which parts to go to last. I know the marshier areas with thick clay soils, where plants may thrive in a drought year, and the better-drained ones where too long without rain can cause them to shrivel and die. I can find the buds of wild maple trees and the vibrant green leaves of the first wood sorrel of the spring. I know that once wild garlic is in flower, before they become seedpods, the rising temperatures will turn the flowers into capers with an intense taste. I know when to harvest the salt-loving wild plants of the seashore and the rocky coasts, and where to find the best almonds, sweet chestnuts, wild cobnuts and walnuts in the country. I know how to preserve, prepare and cook the wild ingredients I find to extract the maximum flavour from them, and how to combine botanicals to produce delectable syrups or distinctive spirits.

I've learned my craft partly from what Grandad taught me all those years ago. Having a grounding, knowing all the plants and the trees in the woods and being able to recognise the leaves, flowers and fruits, was crucial. I've built on that base knowledge, partly from my research but mostly from just doing it. I taste everything I sell and I only sell the things that I like, because if I don't, I'm not going to be able to convince my chefs to try them. I've eaten the needles, the tips and cones of pine trees, for example, so I can offer them with confidence, because I know they taste good – and I'm not going to poison them.

I've been a forager since the year dot, but after first realising I could make money from it fifteen years ago, its been my job for the last ten. No one's ever going to become a millionaire by picking mushrooms and weeds but it is a great way of life, though not everybody quite gets it. Even when I try to explain it, they tend to say, 'What? You pick weeds for a living? How does that work?'

As I've learned for myself, you can make a living while continuing to develop your knowledge of wild plants – and it's now more important than ever that you pass on that knowledge to other people, because otherwise it will be lost. So the next generations of my family are going to learn it whether they like it or not. But I never take my chefs out foraging with me. I don't care how many Michelin stars they've got, that is just not happening. Some of them have pleaded to come, they don't care if I threaten to blindfold them and confiscate their mobile phones, they still want to come, but

I've always managed to worm myself out of making a firm commitment.

Chris and I do not have children of our own, but I've always tried to involve my nephews, Alex and Ryan, in the foraging we do. Whenever they do come out with us, I'm gradually passing my knowledge on to them, just as Grandad did to his children and grandchildren, so that, with luck, the family tradition we all embody will live on through them. Being teenagers and Xbox aficionados, plants and mushrooms don't quite grab their undivided attention yet, but when I sent young Ryan out into a wood recently, he rang me up to tell me he was buzzing because he had found some oyster mushrooms on a dead tree. That little spark of enthusiasm is enough for now, and when they get older, they might just realise that picking weeds is actually a little bit of genius if you have young legs and half a brain, as well as a well-connected aunty.

I pick over a hundred different types of wild food for the chefs I supply, working through the seasons and knowing that the natural cycle doesn't stop for anybody. I'm out nine months a year, in all temperatures and all weathers. I've earned my stripes, got out there and gained the knowledge, and I've learned an awful lot along the way, but I'm always aware that there's still plenty more to discover. When you do it day in and out, you realise that every day is a school day, and you never stop learning.

I'm very lucky that I've got Chris and Mum alongside me, because getting the right partnership is the key to success in work, just as it is in life. And I can rope in my brother and his sons from time to time; Adrian still loves being outdoors, just as he did when we were kids.

It's hard graft, but when I'm out in the woods, on the moors, or at the coast, or sitting in my foraging shed at the bottom of my garden, sorting plants, herbs and mushrooms to deliver to my chefs, I feel a satisfaction that nothing else has ever given me. It's the missing piece of the puzzle I've been looking for all my life.

Witnessing the cycle of life through the flora and fauna I find is good for my soul, and the patterns I see around me have made me more aware of my own place in the world. I'm in charge of my own destiny, and when you're at one with nature, as I am, you can become the person you are meant to be.

PART TWO

My Foraging Year

Yorkshire is bursting at the brim with wild edibles and I want to introduce you to some that might be familiar to you, and some that probably won't be. As a forager, it's not enough just to be able to identify plants, you need to make them your friends with a shared history and a future together. So in this next part of my foraging journey, I want you to learn more about some of the plants we forage, and how our lives have been intertwined for thousands of years, most of which we've forgotten.

Wild Garlic

I always think of spring as the 'green season', and it kicks off with our old friends the alliums. Wild garlic (also known as ramsons) starts growing later the further north you are, by two weeks for every degree of latitude – which means January in Cornwall and mid-February in Norfolk, March in Yorkshire and God knows when in the north of Scotland, unless you're alongside the warm waters of the Gulf Stream. The different growing seasons can be an advantage, of course. As part of a network of trusted foragers, if I need wild garlic when it's out of season in my back yard, I can usually get some from elsewhere and then return the favour in a couple of months' time.

There was an old tale that the Romans planted wild garlic alongside the Great North Road, which followed the same route from London north as the A1, and that the wild garlic in the surrounding Yorkshire woodlands are remnants of a strategy to feed the troops as they were on the march. It's an ingenious idea and could be true, but I'm not so sure that the wild garlic in the woods comes from those same plants.

All plants have different triggers and strategies to survive. Wild garlic doesn't care about snow, frost or cold; all it cares about is light, and when it starts to get enough of it, from around the spring equinox onwards, it will grow like mad. We begin by picking at the top of the wood, where the sun first hits it, and work our way lower over successive weeks, as the light strengthens further down the hillside.

Wild Garlic

Picking wild garlic is not just good for the bank balance, it's very good for the complexion too, because the volatile acids that wild garlic releases as you cut it are like a chemical peel and will eventually strip the outer layer of skin from your face and arms, but in a gentle, therapeutic way, not like you've been flayed alive.

While wild garlic is incredibly common and in some places will cover an entire woodland in a carpet of pungent, green leaves, it's not without its dangers. At the same time as wild garlic is coming through there are three very poisonous plants that like to share the same space and look very similar when all the plants are young. We have the glossy green leaves of lords and ladies and jack-in-the-pulpit, as well as dog mercury, which is shorter than the garlic and hides at the base of the garlic among the leaves. All three are particularly nasty and we have to pay an awful lot of attention, because after an hour or so of garlic picking, when you blink you still see an imprint on your eyelids of what you're picking – being 'garlic blind' is akin to being snow blind.

My chefs can't get enough wild garlic because it has an intensity of flavour that cultivated garlic can't match. So in the season we're picking as much as a tonne a week and I have to rope my entire family in to help, starting at six in the morning and finishing at dead on eleven a.m. A huge problem with picking wild garlic in bulk is that by late morning, not only do the bugs wake up but the sun begins to raise the core temperature of the garlic leaves you've picked. If you have the garlic in five-kilo bags then the middle can get really warm and the garlic even takes on a self-heating life of its own, not unlike a compost heap. So the key is to pick it and get it to the huge industrial fridges of the commercial suppliers like Oliver Kays or Freshview, which are literally bigger than my street of houses. Only then can I breathe a sigh of relief. There is nothing worse than having all your hard work ruined.

For jobs like this, I could hire extra pairs of hands of course, but I've had one or two unfortunate experiences working with other people, mainly lack of experience and not paying enough attention to detail, which can sometimes be the difference between life and death. No, really.

I always have to double-check everything, which is one of the main reasons I don't want to expand beyond what those close to me can cope with. You have to be careful who you work with, because you're giving your secrets away – another reason why we stick to family and are very particular about who we supply. We can't be all things to all people; I want to be all things to a very chosen few.

Unlike most of the other plants we forage, we're rarely troubled by insects or other creatures when we're picking wild garlic because hardly any of them will go anywhere near it. Foxes, badgers and rabbits all avoid it, and when you see pristine-looking plants, with no animal tracks running through them and no burrows, or signs of the foliage being browsed, then you know you're not going to be depriving the wildlife if you pick it. Almost the only creatures that will eat wild garlic are cows, which is not good news for farmers. If they're beef cattle it's not so bad because most of us like a bit of garlic with our meat, but if they're dairy cows, forget about it, because I guarantee that garlic milk is something that is never going to catch on.

We also collect wild garlic capers, the young seed pods, although there is only a tiny window when they are available. The capers are well worth gathering, because a single tiny little bud has the same intensity as a whole clove of garlic.

The trick is to pick the capers when they're a dull green colour, right after the flowers have died and before they go yellow, because by then the seeds will have formed and the flavour will no longer burst in your mouth. I also love the wild garlic flowers and the 'scapes' – the point where the flowers are still encased in a sheath – are a particular treat. Tempura batter them with plenty of seasoning and you'll be fighting over them – they are that good.

When you get towards the end of the wild garlic season, while most of it is still perfectly good and usable, one or two of the plants will start to go over. The leaves begin to go ragged at the ends, turn limp and yellow, and if it starts to rain, they just flop to the ground. When that happens, as the plant begins to decompose and its cellular structure breaks down, it releases all the acids that were stored within it and if you don't remove those dying leaves from among the ones that you're picking and supplying to chefs, they will speed up the decomposition of the rest and reduce what should be a storage life of a couple of weeks to a handful of days.

Strangely enough, wild garlic bulbs cannot be taken because they are on the DEFRA (Department for Food, Environment and Rural Affairs) protected species list; you can pick the leaves, but you can't dig up the bulbs. That feels a bit odd to me, because if you looked at the woods round us, you wouldn't imagine that wild garlic was in need of any protection – it's absolutely everywhere. So it's not one of those plants I'd consider endangered, but just the same, I don't take the bulbs, though even without them, garlic still gives us four crops: leaves, scapes, flowers and capers.

*

Nettles

You'll have noticed that spring encourages young nettles to thrive, but you don't have to see them as the enemy. They're chock full of vitamins and iron, and great as a soup or as a spinach-like vegetable.

Andrew Pern and I did a foraging night at his restaurant, Mr P's Tavern in York, a while ago, when his take on nettle risotto with spelt, cheese and wild St George's mushrooms was a bit of a revelation to me. It's now a favourite dish of mine, always provided I don't have to cook it myself; partly because when you have a Michelin-starred chef cooking your food it really brings the best out in terms of flavour and partly because of bone idleness on my behalf. Nettle leaves are also used to cover Cornish Yarg. It's a very traditional way of preserving; they form a protective barrier and lend the cheese a distinctive taste. Nettle pesto is beautiful when run through dishes that require a bit of extra oomph. Even when the nettle leaves get huge they are great for preserving, and because of their fibrous nature you can parcel them up and cook in them too, in the same way as you do with banana leaves.

Most plants are nutrient sappers, but nettles are one of the very few plants that give back to the earth more than they take out of it – they put a lot of nitrates into the soil. Nettles and puffballs also go together really well; you'll often find that if you come across a huge patch of nettles, it'll be really good conditions for growing puffballs. It's all about the pH value of the soil and what the plants need to thrive, but for a forager, being able to read the landscape is also key. Things grow under tree canopies or in the open for a reason; if a plant is in a particular kind of location, terrain and habitat, it's because those are the best conditions to bring that plant into maturity to reproduce. The by-product of that is they are the perfect conditions to produce the best flavour in the plant.

Stinging Nettle

The other great thing about nettles is that they are cut and come again. Which means we can carefully tend a patch and ensure we have a really good supply of fresh young nettle leaves all year round. It takes a while to get to know which wild plants will come back again and which ones will disappear altogether if you over-pick.

We spend a lot of hours in the foraging shed snipping off the leaves of nettles. They need really careful preparation because they are an absolute magnet for butterfly eggs – in fact anything that needs to lay an egg on a leaf really. My chefs like everything prepped as much as possible, so we spend hours carefully inspecting each leaf for eggs, holes, frostbite, and they have to be absolutely perfect. And bearing in mind that back in the day they used nettles for rope making, we need to make sure that we're getting the chefs the greenest, freshest, most succulent leaves that they are looking for to enhance their dish.

However, it doesn't matter how careful you are with stinging nettles, you will get stung. A draught came through the door last year and we had a full bowl of nettle leaves that blew over Ryan and of course he just had a T-shirt on. He didn't stop moaning about it all day. Of course I did tell him to wear long sleeves but they know better at that age. The interesting thing I have noticed about nettles is that over time I have become nearly immune to the sting. The little glass-like needles full of nettle venom that puncture the skin are at first a nightmare, but I've learned that the trick is to resist temptation and not rub it. An antihistamine dock leaf can help too, or these days I carry antihistamine cream with me.

There are nettles that don't sting though, although people don't often realise this. They are – not surprisingly – known as dead-nettles, and are just as flavoursome as the ones that do. Bumble bees absolutely love the flowers of dead-nettles, and they're not wrong. One variety has tiny red-purple blossoms that are probably the sweetest I have come across – you can pop a few on your tongue and you'll get little bursts of liquorice – while those of the white ones taste of honey.

Waving a bunch of dead-nettles at a chef and seeing him flinch still makes me laugh when he doesn't realise that they won't sting. I've yet to do that to young Ryan because I keep forgetting – next year for definite.

Wood Sorrel

The first wood sorrel leaves begin to show in March, and it's a wonderful little plant. It's absolutely minuscule but packs a really lemony zing when you eat it and it's a huge favourite with chefs.

Chris was over Barnsley way one day, where we have a lovely big patch of it down a dell in the woods. He was busy picking it when he heard a voice above him. 'Nah then, what's tha doin' down there?'

Chris looked up and met the gaze of a couple of old boys, probably retired miners from one of the hundreds of pit villages in the area, staring down at him. 'I'm picking sorrel,' he said.

'Soil? What's tha picking soil for?'

'No, not soil, wood sorrel.'

The two old guys looked at each other, and burst out laughing. 'Wood soil? Why, does it give thee a hard on?'

'I wish it did!' Chris said. They eventually sorted out the confusion and after a good laugh went their separate ways.

There are quite a wide variety of sorrels: we get common sorrel and wood sorrel in abundance round our parts, though I have to say not as abundantly as in Scotland. The early shoots are bright green and make the most beautiful sorbet with the oxalis acid they contain. It's similar to the acid within rhubarb, giving you that lemony sour taste. The plants are only about an inch tall, so they're a huge amount of effort to collect and because they're so low to the ground you have to be careful where you pick them – picking by paths is an absolute no-no. But my chefs love the prettiness of the leaf and the delicate white flower. Unfortunately they pull the stalks off, which I think is sacrilege – the purple stalk is the best bit. And not to mention the cost!

On my travels, I've seen wood sorrel growing in carpets covering woodland floors and I've daydreamed of getting to work with a lawnmower and filling the grass box with wood sorrel in seconds, but the reality is that we're on our hands and knees for hours on end picking the stuff – it's a bit like digging nits out of kids' hair without even having a nit comb.

Field Maple

You have to be quick when you're picking the young green shoots because the seasons never stand still for an instant; they're always evolving, always changing. Every single week is different, so you've always got to be ahead of the curve, but it's taken me a lot of years to reach this point. When I first started as a professional forager, I felt I was always struggling to catch up. There were so many new things to learn. While I was doing one thing or picking one kind of plant, I'd suddenly look round and think, 'Gosh, has that other one come out already?' Every week, every season and every year, I learned a little bit more, and now I feel we can keep pace with the seasons, no matter what the British climate throws at us.

A lot of plants only have a two- or three-week window in which you can pick them; miss that and you'll be waiting another twelve months for it to come round again. So, along with everything else, I always need to be aware of when each of them is going to be ready and at its best

One example of this is the field maple. We pick the shoots in the spring and they are really delicious – nutty-tasting as well as sweet – but two weeks later they've opened out and turned into awful flappy leaves, useful for little more than wrapping food for slow cooking. The leaves of the field maple look like sycamore when they open, and although they aren't as big and robust as the leaves of vines and banana plants, which people have traditionally cooked with in hotter climates, they will do the job. On the other hand, the shoots make a great mixed salad when chopped up and combined with hawthorn shoots, wood sorrel, chickweed and cherry blossom flowers.

When picking, you need to take great care in recognising your trees, as the field maple is a tree that hides among hedges and its young shoots can look pretty much like anything. It's better to identify the tree from the year before, just to make sure, and keep an eye out for those first shoots of spring, where the leaves are wrinkled and curled up and are packed full of sugar, without the chlorophyll taste that they get later on.

You pay the price for that sugar later in the year, when the trees drop all that sticky gunk on your car, without so much as a by-your-leave. We always get charged more at the car wash once they've seen the roof. Their lips purse and the big, heavy-duty wash brush comes out and costs Chris an extra two quid. I don't go myself because I always get charged even more – I don't know why, but he always gets a better deal than I do.

*

Birch

I can only get birch tree sap for about two weeks – in the north, that's usually the second and third weeks of March. By the second or third week of March, the birch tree says to itself, 'Right, I need all my leaves out,' and it's as if it had planted a big straw in a pool of water and started sucking on it. The tree draws water up through its roots and has its own filtration system in the bark, spreading nutrients through the rest of the tree, like the arteries and veins moving blood around your body.

There's always a bit of a wow factor if I am out with someone and cut into the bark of the tree because of the amount of sap that pours out. I always encourage whoever I am with to taste it. It's not hugely concentrated, but the taste is of faintly sugary water and is quite nice.

The sap is really good for you and very high in natural sugar, but only has a twenty-four-hour shelf life and since it's sucking up water direct from the earth, it needs to be pasteurised, which is easy enough to do but a lot of people don't bother.

You can separate the sugars and the water with a centrifuge – I nearly bought a huge scientific one on eBay once, but it was the size of an industrial washing machine so thank God I was outbid. There is the Birch Syrup Company, who have acres of birch in the highlands of Scotland, and they produce an intense pure birch sap syrup which surprisingly resembles barbecue sauce with a hint of smokiness. I have reduced birch tree sap on the hob, but it's a tedious, time-consuming process and not very environmentally friendly because it takes a lot of fuel to turn about eight litres of sap into half a litre of concentrated liquid. I've often relied on the hood of a Gore-Tex coat to do the job instead. Because of the relative density of the sugars and the water, it draws one away from the other in a kind of reverse osmosis. It's a bit simpler than boiling it up all day – and a whole lot cheaper.

To get the sap you need to leave your tap in the tree, even though leaving your equipment out in the woods is always a bit of a risk. You don't see anybody for days but as soon as you leave your gear out it disappears, so these days it's a case of going out into the deepest, darkest, most inaccessible parts of the woods to stand the best chance of not getting robbed. If you manage this, you'll get about a litre an hour. Most chefs want it pure and fresh from the tree, but because I can never tell if my equipment has been snaffled I have to keep it on my 'off list' – meaning I don't put it out there generally – as I can't be sure I will be able to supply it.

Cleavers

Some plants taste great one week and horrendous the next. Cleavers are a case in point. They're the things that wrap themselves around your feet when you're walking and cover your clothes in small burrs, hence their nickname: sticky buds. They belong to the same family as coffee plants, though they contain significantly less caffeine. Cleavers are hugely common. They are probably growing in your garden right this minute, and if not, they will be trying to.

The burrs can be dried, roasted and used as a coffee substitute, if you have the patience to pick them – because they are really quite small – but I concentrate on their new leaves and stems. When they're new, they taste exactly like young, fresh peas and can be eaten raw or cooked. Just blanch them and purée them; they're not only full of flavour but very good for you. I'm quite partial to a cleaver mousse with a bit of smoked fish or, mixed with some white pepper and olive oil, they make a fabulous mushy pea style pesto. One of my chefs does a lobster and cleaver dish, and it is phenomenal. But beware, when they're only a few days old, they suddenly taste of nothing, and become woody and stringy as well.

I often take the very young plants to chefs who haven't tried them. They know immediately what they are, and when they nibble on them raw, they are always amazed to find how much they resemble peas. I'm always astonished at the amount of plants that taste of something really familiar while looking nothing like it. The only trouble with cleavers from my point of view is that, once chefs realise they are a really good foraged food, it's easy enough for them to go and pick some themselves, which is great for them, but not very good for my business.

Jew's Ears

People tend to think of mushrooms coming through in the autumn, but there are a few mushrooms that we pick in the spring.

The first mushrooms of the year usually appear in groups, in damp conditions, on dead and dying branches, almost exclusively of the elder tree. Their name – Jew's ears – is supposedly linked to Judas Iscariot, who Christian tradition claims hanged himself from an elder tree after betraying Jesus, and 'Judas' was mistranslated into English as 'Jew'. I don't know if he really had large fleshy ears, but that's what the fungi look like, and it's no accident they are also known as jelly ears. When fresh, they are gelatinous in nature and quite heavy. In folk medicine they were used as a remedy for jaundice, sore eyes and sore throats, but modern scientific research suggests they have properties that may lower cholesterol, work as an anti-coagulant, and counter hypoglycaemia and some cancers.

The Jew's ear is a parasitic mushroom, meaning it infects its host with spores and congregates in its every nook and cranny. Give them a good rainfall and they grow like the clappers through the spring and even into the summer, looking like miniature brown paddling pools or mouse ears. I find collecting them very therapeutic. I like the feel of them in my fingers and they are easy to prise off the tree, so it's not long before I have a bag full. Where we pick wild garlic there is an abundance of Jews' ears growing on elder, and it's hard for me to concentrate on picking wild garlic when every time I look up I see big juicy mushrooms.

They really are one of nature's miracles. They're the most versatile of edible mushrooms – great stir-fried or shredded and are very popular in Asian cuisine. One of my chefs, Alisdair Brooke-Taylor at the Moorcock Inn at Norland, made a marmalade from Jews' ears, which I loved. I ate it with a pig's head terrine with hogweed seeds which was a whole lot better than that might sound – absolutely lovely in fact. I was surprised at the texture of the mushroom marmalade and the flavour was remarkably good.

You can also dry them to within an inch of their life, then

grind them into powder and use them as a mushroom stock cube or sprinkle them on a casserole dish – it's like instant gravy granules for mushroom lovers, veggies and vegans. So they're a pretty useful mushroom and their popularity is going mainstream. I've even seen them in the speciality section of supermarkets, but there is nothing like picking your own and drying them or making something with them.

St George's Mushroom

Late April is marked by the appearance of St George's mushrooms, so named because it's around the 23rd – St George's Day – that they poke their noses up through the soil and leaf litter. I love the way plants, fungi, berries and nuts sometimes take their names from history. In Yorkshire we tend to have to wait a couple of weeks longer, but when they do come out it can be an amazing sight. Some years they fill the woods as far as the eye can see. They tend to grow in troops, and I will always pick the first four of a troop and leave the rest alone to spore. Chefs are particular when it comes to sizes, you need small ones for plates, so taking the first four is not a bad thing because towards the end of the troop they become enormous. When I give them a clean-up at home, I also take any mycelium – root – that's still attached and put it somewhere I think is suitable to give it a chance to spore. This technique worked very well with the field blewit mycelia I scattered in my garden; I now have a bumper crop every year.

At Moor Hall Mark Birchall puts St George's mushrooms with Herdwick lamb, crispy lamb belly and roasted asparagus – a concert of spring flavours, with a dash of anchovy that you wouldn't think would work but that actually lifts the dish to another level. To be fair, Mark could elevate anything to a different level; it takes his team three days just to make laminated onion bread. It's a work of art and I'm always on the cadge to take some home . . .

St George's Mushroom

St George's are one of my favourite mushrooms; they have a great fridge life and are the chunky monkey of mushrooms as they are so heavy for their size. They have a fabulous aroma too, a bit like flour. If you know a certain spot for them, always go looking five days after a rainfall in April and you should be in luck. I always find mine in woods, usually on pine litter, although you can actually find them out in fields, or on the edge of woods in a clearing . . .

I take a bright-coloured rucksack with me and hang it on a branch, because it is so engrossing picking St George's that the hunt takes you deep into the woods and by the time you look up, you're very much off path. A little wander and a pretty good sense of direction will usually get me back to my brightly-coloured rucksack hanging in a tree. You need to work methodically too, by ignoring other side paths till you have worked a certain area, or you will be all over the place and repeating the same ground. Picking these mushrooms is one of my favourite things to do, but they do get dirty really quickly and the gills are magnets for bits of muck. If you're not sure where to start, look along the paths in your local wood. All mushrooms are big fans of living near paths and you'll find the St George's hidden between two to ten feet away from the path.

Dryad's Saddles

At the same time as St George's, dryad's saddles are also coming through. Other common names for them are devil's saddle and pheasant's back, because the brown-flecked cap looks a bit like their feathers. They're a parasitic mushroom, a polypore or bracket fungus, which you find growing on the old stumps or sometimes the trunks of beeches, oaks and other trees, and they can be really massive, the largest capped mushrooms in the UK. They have a beautiful dogtooth check pattern on the top and are a creamy white underneath. They grow to huge sizes and literally look like saddles. I've found some that were two feet across, but by that stage they're as tough as old boots.

When they're young and tender you can just pick, chop and fry them, and they have a really great taste. However, they seem to come and go; one year they're everywhere and the next there's not a trace of them, but in good years every tree stump will suddenly seem to be dripping with them. The key is to look under the big brackets to find the smaller ones that are protected as they come through. Lift the largest of the mushrooms and there will be a whole host of small tender dryad's saddles trying to come through, and these are the prized specimens that you want. They are a really underrated mushroom, and I think it's because most people are more excited by the large ones whereas it's actually the teeny tiny ones that are the delicacy.

Gorse

If you're going to pick gorse, you've got to make sure you can tell it from broom. If you don't know what you're looking for, they can be hard to distinguish, as the plants look similar to an inexpert eye and broom is not on my hit list of chef's ingredients. Gorse has flowers and young green shoots that smell and taste like coconut, and make a great infusion. And while broom has the same bright yellow-looking flowers of a similar shape, it's the coconut smell of gorse that gives it away.

Gorse blooms twice a year, once in spring around May time and another half-hearted attempt at the beginning of October. I say half-hearted because only the tips seem to come out, whereas the spring bloom is the whole bush.

If I need gorse when I'm back in Yorkshire, I usually pick it up by Conisbrough Castle. It really is pretty there and is one of the few places around Doncaster that gives you a good vantage point over the area. So I can enjoy the view and also fill a bag with gorse shoots that have that wonderful flavour.

Gorse is hellish to pick because the needles are really sharp and spiky. However, no pain, no gain, and anyway my fingers are quite nimble so I can work my way fairly deftly around the spikes. When we pick gorse we wear Gore-Tex jackets, as they're the only things that will not get you entrapped in the gorse bush. Gore-Tex is like a prickle repellant. If you wore a fleece it would be like wearing a big piece of gorse Velcro, and once you get stuck it's near impossible to get out of. There was a chap a few years back who needed the Air Rescue helicopter to come and pluck him out of a particularly thick patch when no one else could get to him.

Gorse

It's all worth it, though, because those coconut-scented flowers are some of the most amazingly flavoured blooms that grow in the wild. When you have picked a good bag full and you give them a shake it smells of pure coconut, which always blows my mind. Sometimes things don't transfer well by infusion, but the gorse flower does and is a real treat. In fact, I'm thinking of doing a distillation of gorse flowers because it's so unusual. My chefs are amazed by the intensity of its scent and taste, and revel in creating something wonderful from it, like gorse flower panna cotta.

Morels

When we get into May, the morels are starting to come out as well.

We send out a fair few mushrooms come the season, so cleaning and prepping them all can be a very long job, and morels always seem to be full of bloody woodlice, which makes them even more a chore, but the end result is well worth the effort, even if they do look a bit like rubber brains. If I didn't clean them, the woodlice would probably add a third to the cost and they aren't the cheapest of mushrooms to start with.

A chap in Wales once delighted in sending me videos of the copious amounts of morels that were growing in his local area. So I persuaded him to send some to me on the assurance that it could be a mutually beneficial arrangement. I spent a fair bit of time telling him how delicate they were and the way they needed to be picked, sorted, cleaned and packed, if they were going to be saleable to restaurants. 'Yeah, yeah,' he said. 'Got it.'

A box then arrived one day, sent through the post as a standard parcel, so it had taken a while to get to me. It wasn't just any old box, it was a really damp and smelly one, the kind that you find at the back of your shed that's been there a few years, got damp, dried out and got damp again, over and over. In this box there were a large quantity of morels that had all been squashed flat in order to squeeze as many as possible into the box. And they were not only squashed, they also smelled strongly, not of morel but of mouldy cardboard box. I had to throw them all away and then break it to him that we weren't going to be a good fit as trading partners.

Morels

If any of the morels or the other edible mushrooms we've gathered aren't fit to be supplied to restaurants, because they've either been squashed, or got at by creatures, or are past their prime and ready to spore, we'll always try to take them back to the wood where we picked them, or to a new wood, and scatter them there, so that they have a chance to spore and regenerate. It's all part of ensuring the continued survival of the wild plants we pick and the biodiversity of the wild areas we visit. I can't stress enough how important it is for us to look after what feeds us.

*

Elder

The elder tree is the gift that keeps on giving for foragers, because it provides three different crops a year: first flowers, then capers and finally berries. The only part of the elder that is pretty much useless to a forager is the wood. The branches are not particularly pleasant-smelling and they're quite soft and hollow in the middle, so they're useless as fuel, though blowing through the hollow stems has the same effect as using bellows to fan the flames of a fire, leading some people to believe the name elder derives from the Anglo-Saxon word for fire.

Elderflowers are beautifully scented, but they're always crawling with bugs and pollen beetles, and black-fly as well if you pick them in a garden, unless you do it really early in the morning. Like any flower you fancy picking, the brighter the weather, the more bugs you will get; so on a cloudy day, you will avoid the lion's share of them. Whatever flowers use to attract pollinating insects – not simply the scent, but their intense colours, right across the spectrum (including those that the human eye can't see) – is diminished when the sun isn't shining. A hot tip is don't wear white or yellow when picking any blossom in the sunshine. You will quickly become covered in pollen bugs and the blighters don't leave you alone.

When I get the elderflower heads home, I put them into a plastic garden refuse sack and a box with a small hole in the top, and leave it in the sun for an hour. The creepy-crawlies are so hot by then that they leg it as fast as possible for the hole, desperate to escape. I can then put the bug-free elderflowers somewhere cool. It's the kind of thing you learn from experience – in my case after the inside of your car has become so smothered in pollen beetles that you can't even see out of the windscreen.

Elderflower

I love the smell of elderflower, and the lower part of the flower-head is just as fragrant as the top, but sadly it doesn't dry well. It has a fairly robust stem, so doesn't wilt too much, but while you can hang meadowsweet up to dry without the petals dropping off the stem, elderflower just disintegrates. The flowers drop off, and unless you've got them spread out on an enormous tarpaulin, they tend to form a clump which you then can't separate. My chefs use elderflower for savoury and sweet dishes. I've seen some really imaginative pairings with meat and fish, as well as elderflower desserts such as 'champagne' and panna cotta. I've even seen elderflower heads coated in tempura batter and lightly dusted with sugar.

Elderflower capers are good too, very like the capers you buy in jars from delicatessens and supermarkets, but with a rather more floral tone. However, the raw fruits of the elder tree must be cooked or infused before being eaten because they're mildly toxic if not, and may give you a bit of a dicky tummy.

Come autumn and it's time to harvest the elderberries. When the stalks of the elder have turned a shade of purple to match the berries it's the perfect time to pick them – they want to be full of juice and lush, with a burst of flavour in the mouth. It's not a berry that you really want to eat on its own. It's not like a strawberry or a raspberry, it's a berry that needs to be processed before it is absolutely edible, but the results can be stunning. Elderberry and sloe syrup is magic, and I'm personally very partial to elderberry vinegar. It's just so good and versatile that I make a couple of batches every year to last me through winter. No point in being a forager if you don't have your own personal stash of things.

It's always wise to remember that when picking elder flowers, capers or berries, just like any other plant, you should only take a few from each tree and then you won't harm it. It's easy for me to stick to that because at five foot four I can never reach the medium to high ones anyway, but if you're six foot odd you could have a real crack at it – but please don't.

Watercress

Wild watercress is common in fast-flowing streams, but should only be picked in areas to which sheep and cattle have had no access. You have to remember that water plants are like combs and they try to collect everything that flows through them. Parasites that live naturally in our streams, rivers and lakes are easily caught, and the small eggs are not something you want to be consuming. We don't supply it, just in case, because of the risk of liver flukes. If you eat watercress or freshwater fish containing the eggs of liver flukes, or drink contaminated water, they migrate to your liver and can live there for decades, causing fever, jaundice and anaemia, and a whole lot worse.

I get a bit pedantic about food safety, as there isn't a lot of knowledge out there on how to process things we forage, and there's a lack of general awareness of what we need to be careful about when using wild food. My chefs can't always be aware of what needs to be done, so where I come in is by not only supplying them with great produce, but with my knowledge too.

Watercress has a great peppery taste and is really good for you, so although I don't supply it to chefs, I do collect and eat wild watercress myself because I can guarantee that I have given it a really good clean. I don't eat it raw, but a nice home-made watercress soup ensures that the leaves are heated to a point where anything is killed.

Stonecrop

A great many of the green leaves we pick have gone over before high summer, but at that point, stonecrop is just coming into its own. It's a lime-loving succulent, a bit like an alpine, and grows in crevices like the limestone pavements that surround Yorkshire's Three Peaks, or the limestone cliffs at Malham Cove. The waterfall that roared over the Cove in ancient times must have rivalled Niagara, but, as is always the case in limestone country, the river slowly dissolved the rock and created new underground watercourses. That process created the potholes and caves that honeycomb the Dales, but has left Malham Cove dry except in unusually wet spells, when for a brief period water once more tumbles over the lip.

The stonecrop that grows there has tiny little leaves that look like a big fat grain of rice. If you pop one into your mouth it has quite a nice lemony flavour. The only thing I can compare it with is the backside of a frozen wood ant. As I am sure most, if not the vast majority, of those reading this will not have tried a frozen wood ant bum, it's a bit difficult to describe, but it definitely has a lemony taste. I find all stonecrops are actually best pickled and are like miniature gherkins with a kick.

We also have roseroot, which is another variety of stonecrop and quite common around the limestone outcrops of West Yorkshire. It smells of roses but doesn't taste like them. You have to watch out because there are lots of different types of stonecrop – some purple, some green – and they can be quite bitter and sour. One succulent leaf of the peppery ones can easily burn my throat and I wonder why I put myself through the torture sometimes. But nothing ventured, nothing gained, literally in my case.

Dandelion

The one thing about dandelions is that they really do try hard to sprout all year long. The slightest sign of decent weather and they are growing like crazy. We've all got dandelions galore in our back gardens. They're one weed you can't get rid of no matter how hard you try. It seems as if their sole purpose in life is to sprout up all over your garden, but you can make use of them.

Dandelion

Most people know you can eat the young leaves. They make a great base for a foraged salad – just don't gobble up too many of them, though, because they have diuretic properties; if you overdo it, you may discover why the French call them *pissenlit* – pee the bed. We pick a lot of dandelion leaves, and if you stick a bucket over a patch so they stop producing chlorophyll and go pale yellow, you can charge twice as much for them. It takes the excessive greenness out of the flavour so they have a more subtle, 'stripped down' dandelion taste. You can eat the flowers too. I've had chefs desperate for dandelion flowers in winter, and sometimes you can still find them if it hasn't been too frosty for a few weeks.

Dandelions also have pink roots which taste a bit like chicory or endive. If you dig them out of your lawn, they probably won't be any bigger than the thickness of your finger, but in places where they're well established and tightly packed, they can be as much as four inches across. Like all roots, they are always best come winter. This is because when the plant isn't trying so hard to grow, a lot of nutrients are transferred back down to the root in preparation for the next spot of decent weather. You need to use them quickly, though, as they go soft in twenty-four hours, but sliced with a mandolin and fried in hot oil, dandelion crisps are fabulous.

While we were picking dandelions a while ago, so many passers-by kept stopping to ask us why we were doing it that in the end I started saying that we owned a rabbit sanctuary. I nearly had some baseball hats made with 'Rabbit Sanctuary' emblazoned on them, but that didn't seem such a good idea when people started offering us pets they didn't want any more. Since rabbits breed like, well, rabbits, it could soon have got out of hand.

Horseradish

As a general rule, it really isn't a good idea to dig up the roots of lots of wild plants. It can be more trouble than it's worth, and you can give yourself a hernia in the process because so many seem to love ground that's as hard as concrete. You might as well go and buy them. But I'm prepared to get a hernia for horseradish because I really love the stuff.

The wild plant is a good bit hotter than shop-bought. The intensity of the oils and the strength of flavour and heat you get from wild horseradish – and there's tons growing around us – is infinitely greater than from the cultivated plant, because the wild one has naturally found the best spot for it to grow. But you need to approach it with care. Once you disturb its home, it spreads and spreads, and is nigh on impossible to eradicate. If you drop a piece the size of your thumb in the grass it will grow a whole new plant. I have a healthy recurring crop of wild horseradish in my garden, but it wasn't intentional, it just happened that way.

Horseradish

The plants look like giant dock leaves in summer and can grow more than three feet high. I usually take a picture of them then, because the best time to dig them up is when the leaves have withered and the nutrients have been pushed back down into the roots, but by that point they're impossible to find if you don't already know their exact location. I spent years thinking it would be easy to come back to them in winter, but there's absolutely no trace of them then, so taking pictures in summer is definitely the way forward.

I love taking horseradish to chefs, especially when it's been dug fresh that morning. Luke at Jöro in Sheffield has his team on standby, I come in with about five kilos and they are waiting with goggles and gloves on ready to prep and shred it. I personally wouldn't like to prep five kilos of horseradish, it's bad enough digging it up and that's in the open air, never mind a small kitchen area.

When you prepare roots like horseradish, cutting into them causes them to release oils. It's a defence mechanism by the plant to try and put you off eating it, not that it's ever going to work on me, because I love horseradish, and the hotter the better. The horseradish we gather is really fiery, but you need to use or preserve it in vinegar or oils within thirty minutes of cutting it because it's really volatile. That puts it into suspended animation and it will then keep for months.

It doesn't keep forever though, as one chef I supplied has probably discovered by now. He'd just taken over at a spanking new gastro pub when he gave me his first order: fifteen kilos of wild horseradish. His restaurant had about sixty covers, which meant that even if each serving was a generous five grams, he'd still have to sell three thousand dishes to use it all.

So I was faced with a not entirely unfamiliar dilemma. Should I say, 'What on earth do you want fifteen kilos for?' Or should I just leave him to get on with it? Call me a cynic if you like, but I took the second option, with cash on delivery. You can take the girl out of Yorkshire, but . . . well, you know how that goes.

Burdock

When we were kids the highlight of the week was when the Ben Shaws pop man came round with his flatbed truck with all the bottles tinkling away on the back. It was such a big treat to get bottles of pop delivered. I always went for the Dandelion and Burdock or the Cream Soda, while my brother had the Cherryade. I don't think they ever made it to the fridge – I always remember pop being drunk at room temperature, which usually meant tepid. Best thing was if you took the bottle back to a shop in the village they would give you 10p for it. Best form of recycling ever! Ben Shaws was a local company for us in Calderdale, having started in Huddersfield in 1871, and it's still going strong. Sadly they don't do the luminous, chemical-ridden varieties like Space Special any more, but they still do Dandelion and Burdock and their Bitter Shandy is really good, especially warm with fish and chips.

I never thought about that Ben Shaws Dandelion and Burdock being actually made from dandelion and burdock as a kid, but now that I am older it fascinates me that this must have been a really old-fashioned tonic recipe from generations gone by. The flavour is curious, somehow both bitter and sweet, and with a background taste a bit like an artichoke, which in fact is a close relative.

Burdocks can grow quite tall and bushy, and their roots are ginormous. If you want to go looking for it you just have to know how to distinguish between lesser burdock and greater burdock, which is the one you want. The main difference in the mature plants is that lesser burdock has tightly clustered flowers close to the stems, and greater burdock has flowers that reach out to the furthest stems; so the key is: clustered or not clustered.

But be warned: trying to pull one up is like trying to extract a crampon from a rock face with a cotton bud. You might think you've been clever by waiting until it's rained before getting stuck in, but even the heaviest downpour is only going to penetrate about an inch and a half of soil. Lower down, it'll be rock solid, and a burdock root's grip is tighter than a Yorkshire farmer's hold on his wallet.

Once you've got them, you have to boil the roots for about eight hours to get them into a suitably pliable state. It's a case of: 'Boil them at Easter and they will be ready by Christmas.' It's a bit like cooking brown wild rice. So it's what you might call a time-consuming job, but in less well-supplied times, people needed its sugars and carbohydrates to help them survive over the winter.

But don't let me cramp your foraging style. If you fancy turning back the tide and making your own, you could forage your dandelion and buy your burdock root from Oriental supermarkets. They call it gobo root, and eat it as a vegetable in China, Japan and Korea. It has quite a crisp texture and a sweetish flavour, and you can remove or reduce its slightly muddy undertone by shredding and pre-soaking it in water for a few minutes.

A final warning about burdock: beware of the burrs. They'll grab hold of your clothes, your dog's fur, anything, in fact, that will allow them to reproduce and spread. They're so difficult to dislodge that they're said to have been the inspiration for hook and eye fasteners, and later for Velcro. I've tried washing, freezing and all sorts to try and get rid of them, but nothing seems to work. As I don't want to shove yet another otherwise perfectly good fleece into the recycling bin, I'm sticking with my Ben Shaws.

Wood and Water Avens

Wood avens is also known as herb bennet and is quite a common weed in gardens. The wood avens have pretty little five-petalled yellow blooms and those of water avens are even more beautiful: bell-shaped purple flowers with yellow-orange inners. Wood avens are quite short whereas water avens can grow more than three feet high and have beautiful flowers. You can eat the young leaves of both plants in salads or stews, and the roots have a really distinctive clove scent and taste.

Water Aven

The wood avens' root is only about an inch long, and, like burdock, really hard to dig up, and a bit pointless anyway, whereas the water avens' purple root is six to seven inches long and, growing as it does in the soft mud at the bottom of a stream or pond, is so much easier to get at, but it's definitely a washing-up glove job, because I'm not that keen on the feel of putting my hand into something cold, wet and dark that you just know is full of unmentionables.

Once you've got it, you can dry it out and add it to gin – because it's a root, it's a fixer. You just need to balance it out with other flavours because – just like when you're making a Christmas cake – if you're using several different ingredients you don't want any one of them to be too dominant. It's very good with apple, or for making vinegars. It's also quite useful powdered, as you can substitute it for cloves, giving a bit less overpowering flavour. And back in the day, people also used bits of the dried root as mothballs, and their scent masked the musty smell of unwashed clothes.

I know low ovens and dehydrators are kitchen miracle makers these days, but if you are going to dry your water avens' root it's much better to dry it naturally at room temperature. This ensures you don't destroy the volatile oils that give the root its clove flavour. It will take a few weeks, but the slower you dry roots the better preserved the flavour. And at least it doesn't take as long as the common gin ingredient orris root, which takes five years to get the best stuff.

*

Wild Carrots and Parsnips

There are fashions in fine dining just like everything else, and carrots have been having a bit of a moment recently. I was eating at Moor Hall in Mark Birchall's early days there and one of the dishes was a plate of carrots. You have to be some sort of genius to be confident of serving up a plate of carrots, though these weren't just any old ones. This was the best of British heritage carrots, with every type, shape, colour, temperature and texture. I am not normally that keen on carrots but these were a revelation; they even had cheese snow on top. It wasn't long before almost every restaurant I went to had a version of that plate of carrots and while imitation is the sincerest form of flattery, I was soon wishing that someone would do something spectacular with a plate of parsnips, just for a change.

You can find wild carrots and parsnips easily enough; we have masses of both growing near our Doncaster home. The parsnip plants can be six foot high, with a beautiful yellow umbellifer and really good parsnips on the bottom of them. Wild carrots, which unlike cultivated ones are always white, have small white umbellifers with a tiny purple dot in the middle, and underneath it, a network of spiky leaves that looks a bit like a cage. They started out in Persia about three thousand years ago but didn't stop there. I've holidayed on plenty of Greek islands and found them everywhere. They're edible when young, but rapidly become woody, so the trick is to harvest them while they're still in flower, and to loosen the earth around them first. If you just try and pull them up, you'll leave the tasty bit in the ground and end up with a three-foot stem chock full of phototoxic sap with the little woody inside bit of the carrot on the end.

If you cut the stems or even brush against them, you'll get a red rash and sometimes blisters which may leave dark marks on your skin that will take weeks to fade. They do taste good, though, and you can usually tell by the size of the stalk if the carrot is going to be worth getting. The same goes for parsnips, which also have phototoxic stems, and the phototoxicity is worst when it's really sunny. If you're doing anything with the carrot family, it's best to do it on a cloudy day and always with gloves. It's not often you get a warning about picking parsnips and carrots, but there you go.

When you think of parsnips and carrots that are commercially grown, you think of them being harvested in autumn and winter, especially as they say parsnips are better picked after a first frost so that the sugars in them have been broken down. But wild carrots and parsnips couldn't be more different, and they are at their best in high summer. Once the rain comes and the weather gets a bit manky, so do the roots, as they rot very easily, which has always surprised me. The seeds make for good harvesting too.

The thing to note when working with the carrot family is that for every wonderful, flavoursome plant within the family, there is a doppelgänger that is evil. The deadliest of these is the hemlock water dropwort – it has lovely roots that look beautiful and enticing, but it's not nicknamed dead man's fingers for nothing. So unless you have a bit of a death wish going on, you really have to be sure which is which. Hemlock is deadly, a real cattle and human killer. So if you're ever foraging within the carrot family clan, take a reference book, or better still, take two books, because you really need to be sure with this one.

Hogweed and Angelica

Hogweed and angelica are also members of the carrot family. In the first flush of youth, they are virtually identical, with white cluster umbellifers, and it takes time to develop a keen enough eye to tell them apart. Only when the mature plant flowers and its leaf formations are complete can you tell the difference – hogweed's are shaped like goose-feet while wild angelica's are pinnate – which is on the pointy side if we don't get technical about it.

Hogweed seeds are wonderful when dried and ground into an exotic spice that complements pork and game terrines, and though I won't touch its roots, the French can't get enough of them. I was talking to Mushroom Martin on the phone while he was busy digging up hogweed. He had put on double gloves to protect himself from the photo-toxic sap and said a chef – who was French, of course – had been bugging him for it so he was picking it as a favour. The giant hogweed looks pretty similar to the other members of the family when it starts out, but ends up causing the most damage; its sap is so phototoxic it can cause the equivalent of third degree burns, especially when it's sunny. In a slow summer at least one tabloid newspaper will publish a picture of the giant's latest victim. Animals know they should steer clear of them, and somehow pass this knowledge from generation to generation. I wish they could tell us how they do it.

Animals won't eat wild angelica either, because of its resemblance to hogweed, even though it's not only harmless but delicious too. All parts of the angelica have culinary or medicinal properties, but it's most well-known for its stem, which traditionally has been candied and used as a cake decoration. I remember my mum keeping a tub of them in the pantry which she used to cut into diamonds to decorate a cherry or fairy cake, and I loved the taste – sweet, but with a faintly bitter herbal note coming through, almost like a very sweet piece of celery.

Nowadays angelica has achieved a new popularity as a fixative; the root stabilises the other botanicals in gin. Not only that, but the freshness and flavour of the stem and the seed pods give beautiful bottom, middle and top notes to a gin, creating a wonderful development of flavours.

Pignut

If you're old enough to remember the Calypso lollies you used to get in the 1970s (the orange, pyramid-shaped ones you put in the freezer), pignuts – a cousin of the carrot family – look like smaller versions. Like dandelion roots, they need to be used straight away, because they'll go soft and become inedible within twenty-four hours. There are a number of different varieties, which are now protected by law. I used to dig them up occasionally in the past, but I probably wouldn't bother now, even if it was allowed, because they're swines to get at.

The pignut plant is about six inches high and grows in woodland. It has a white umbellifer, a green stem and little green leaves a bit like fern fronds, but the root can go down about twelve inches and gets thinner and thinner on the way. By the time you get to the actual nuts, it's as fine as a human hair, which makes tracing its path through the soil nigh on impossible. To make your life even more complicated, it never seems to travel in a straight line, so the nuts can be some distance north, south, east or west of the stem.

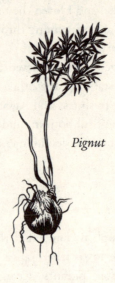

Pignut

If you want to find pignuts – and you're still allowed to on your own land – the easiest way (though easy isn't really the word for it) would be to cut a hole about a foot in diameter around the plant and then rummage around in the soil till you find something – or not, as the case may be. Even then, you'll probably only find a couple of the four or five you're hoping for, because the others are probably hiding in the soil you've extracted.

It's not just digging them that's the problem. Extracting

the pignuts is a two-jumper job even in the heat of summer, because you have to sit in one spot for quite some time getting them out. And as you do, this lets the diligent mozzies sit on you for ages trying to pierce your skin.

Decent-sized nuts are as hot as the fieriest chillies, but it's the kind of heat that sneaks up on you. You'll pop one into your mouth and be thinking, 'Ooh, that's quite nice – a very mild horseradish type taste,' when all of a sudden it'll blow the roof of your mouth off. I'm not particularly good with chillies, so my idea of hot compared to anybody else's is probably like having one drop of Tabasco sauce in a massive pan of chilli. I once took some round to Aiden Byrne, who was quite fascinated with them. As soon as I told him the shelf life, he shouted to another chef to put them on the menu for that night.

It's worth mentioning pignuts even if they are a bit of a foraging nightmare. They are a forager's prize, because there is something significant about being accomplished enough at foraging to be able to get hold of them. Now that I've achieved that, I think I will pass the baton on to young Ryan and see how he gets on picking them.

Sweet Flag

Most of the wild plants we pick are quite easy to find and identify, always provided you know what you're looking for, and have spent virtually your entire life mentally mapping the areas where they grow. But some can be much more elusive. Sweet flag, also known as tangerine root, had me hooked as soon as I read about its intense citrus scent and taste, bizarrely in an old book about seventeenth-century fashions. It was first brought to Britain on the spice trade route from the East Indies, where it had been traded for hundreds of years, and it's been used as a sweetener, an essential oil in the perfume industry, a treatment for various ailments, a crystallised sweet and a flavouring for food, wine, absinthe and even pipe tobacco.

I came across a reference to it when reading about the differences between Coca-Cola and Dr Pepper back in the late nineteenth century when they were both apothecary-type remedies. Coca-Cola had cocaine in it and, as well as sweet flag, Dr Pepper contained a hallucinogenic: mandrake root. So one would get you off your face and one would get you off your head – perfect; no wonder the two companies were in vigorous competition with each other, and both were incredibly popular.

I could buy imported sweet flag fairly easily, but a lot of the flavour and aroma is inevitably lost in the commercial drying process, so I wanted to track it down in the wild. I knew that it grew in marshes, but I'd searched all the swamps, wetlands and nature reserves I could get my hands on, and I couldn't find a single plant.

I needed to think about it logically. When sweet flag was first introduced here, its primary use was as a perfume. That tends to be the British way: if it smells nice, don't eat it – nail it to a tree or your front door to ward off evil spirits, sleep on it, or sniff it.

Sweet flag was used as a scented powder for the flowing 'periwigs' or 'perukes' that men wore in the seventeenth and eighteenth centuries – and not just because they were the height of fashion among the upper crust. Their popularity was partly the result of the spread of syphilis, then known as 'the great pox', which first appeared in Europe in the fifteenth century and reached England early in the sixteenth. Its symptoms included rapid hair loss, which young and not-so-young men were keen to disguise.

The rich could afford wigs of fine human hair; the less wealthy couldn't be as choosy, and sometimes resorted to horse or goat hair, which looked terrible and smelled worse, a bit like wearing a wet dog as a hat. Having said that, even the aristocrats' headgear tended to be pretty stinky, since regular bathing and personal hygiene weren't high on the list of priorities back then – which is where the strongly scented powder came in. It was typically made from jasmine, bergamot, orris root or sweet flag – in other words, a criminal waste of the ingredients for a classic gin recipe.

The 'Sun King', Louis XIV of France, was said to have had a thousand wigs to cover his thinning hair and employed forty wigmakers to create and tend them. As a result, increasingly long and flowing wigs became all the rage, and Charles II, who had fled to the French court during the English Civil War, brought the fashion back to Britain when he returned from exile. They became a whole lot less popular after the French Revolution, when the sight of hundreds of bewigged and powdered French aristocrats going to a hot date with Madame La Guillotine may have persuaded their English cousins to drop the habit in case the same fate befell them – and Parliament's introduction of a stiff tax on wig powder in 1795 finished the job.

Until then, every great house boasted pomanders of tangerine-scented, dried and powdered sweet flag, but it was expensive and not easy to get hold of, so you had to grow it yourself to guarantee a supply. If you were really posh, you'd have a water garden close by your family seat, and as sweet flag is naturally a swamp plant, it grew well in them. It's a member of the iris family and when not in flower closely resembles them at a casual glance, except for a distinctive variation to the edges of their leaves. Other iris plants have one or more serrations on each edge, but sweet flag only has serrations on one side, so I knew that was what I needed to look for.

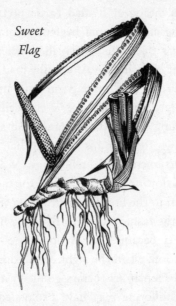

Sweet Flag

I began checking out the water gardens behind our local stately homes, some of which had been pretty much untouched for generations. It took a while, but one day I was walking down a steep hill towards the water gardens of one stately home, admiring the ducks and the swans, when I saw the one-side-only crimping of the leaves of the plants surrounding them. As I looked down the stretch of water, I could see loads of the stuff waving at me, as far as the eye could see, having grown there unnoticed for over three hundred years. I was really excited. If I'd come earlier or later, the flags would either not have been fully formed, or would have gone over, but there they were, cheerfully welcoming me in enormous quantities. So, ferreting out sweet flag's origins, properties, characteristics and historical uses had enabled me to track it down. Without the detective work, I might still have been searching for it to this day.

I become a bit of a giddy kipper when I've managed to capture a plant that has eluded me for so long, but I kept my head long enough to get permission to pick some from the curator of the stately home – and no, I'm not going to tell you which one. He had absolutely no idea it was there; even the head gardener didn't know what it was. When I showed him it, he still didn't have a clue.

You can use the leaves, but it's the root that's unique and you can pick it all the year round, apart from a couple of months in high summer when all its energies are going into leaf production and although useable, it's a bit soft and dingy. When you snap it, you get an overpowering scent of tangerine, but you can't just munch it straight from the pond, nor should you peel it, because the aromatic volatile oils are contained in its outer layers. It's an incredible ingredient, after you've scrubbed it clean and got over the fact that ducks and swans have pooed on it for months on end, it may contain liver flukes, and it smells really badly of swamp. You also need a crowbar to prise it out of the mud while sinking into it yourself and praying to God that you haven't got a hole in your wellies.

My next mission was to learn how to process my booty. It was so incredibly astringent – like an unripe tangerine on steroids – that I had to find a way to reduce that without losing the magnificent taste and aroma that went with it. Trial and error was the order of the day, and it took me ages to get right. Luckily, I'd shared my adventure with a friend, Jess O'Keeffe, who is an amazing chocolatier and pastry chef, and what she doesn't know about desserts isn't worth knowing. So I said to her, 'How can we turn this beautiful stuff into something that tastes as good as it smells? It's so astringent, it's inedible in its natural state.'

'Just give me some,' she answered. 'I'll see what I can do with it.' Jess loves a challenge.

She turned up the next day with some candied tangerine root and some white chocolate with a subtle and distinctive flavour. I tasted some and it was magic. 'You've cracked it!' I said. 'How on earth did you do it?'

She laughed. 'I just boiled it, and boiled it, and boiled it. I've been up all night!'

She boiled the sweet flag root ten times in all, changing the water each time and starting again from cold. Since then, I've had chefs who've just boiled it five or six times and then given up in disgust, thinking that it would never lose its astringent edge, but if you persevere, you end up with something that really does taste as good as it smells.

While Jess was working out how to candy it, I'd discovered that it's also really good in alcohol. It makes a gorgeous root beer and it's brilliant in gin, because of its fixative properties; it stabilises the more volatile ingredients, while imparting that incredible flavour and aroma of tangerine. Combining it with orris root works fabulously with gin, giving a perfect backdrop and base-note. Sometimes life is full of happy coincidences like that, especially where alcohol is concerned. But having the knowledge of botanicals helped too, because I knew, almost like second nature these days, what was likely to go with what.

I took some sweet flag when I went on *James Martin's Saturday Morning* on ITV, in the form of a caramel sauce. He took a tentative taste, then his face lit up and he couldn't stop eating it. In fact after filming he commandeered it for himself. He said he was going to share it with his friends, but my money is on him sitting with a spoon eating it on his own.

Sweet Cicely

I discovered sweet flag after months of sleuthing, and finally twigged that I seemed to have a problem every time I set out to find a plant with a common name beginning with 'Sweet'. Sweet cicely, a member of the wild carrot family, proved to be just as elusive. I'd been scrambling up and down woodland edges, grassy slopes, drainage ditches and stream- and river-banks for ages, risking life and limb to seek it out, but without success. Although I knew what it looked like from book illustrations, I had no actual experience of the plant itself and it was vital to be able to get it right, because while sweet cicely has an incredible fragrance, it's virtually identical to hemlock, which is deadly.

While searching in vain for sweet cicely, I tried a lot of things that tasted bloody awful – but to protect myself, in case I got it wrong and started chomping on hemlock leaves,

I relied on SAS-type survival techniques. If the boys in black find something they think might be edible, they rub it on their lips and wait to see if they get a reaction. Next, they run it around the inside of their mouth – because then it enters the bloodstream – spit it back out and again wait a while to see if there's a reaction. Only then will they swallow a little, and wait to be sure it is not going to make them vomit or worse before really tucking in. I've often had to go through that process when trying to track plants down. If I got a wild carrot stem, for example, and rubbed it on my lips, its phototoxic sap would give me an acid burn, and I'd know at once that it wasn't good to eat. If I put something in my mouth and it started stinging or my mucus membranes started swelling as it invaded my bloodstream, I'd know I was entering very dodgy territory.

I knew I was searching for sweet cicely in the right sorts of areas, but I just couldn't find it. Then one day my mum brought a sprig over to show me. She was all excited and I could tell she was dying to tell me something. Out it came as she wafted a big bunch of green stuff in front of my nose and the smell was unmistakably of aniseed. The look on her face was pure delight. 'I think I've found sweet cicely,' she said.

I was already holding some hemlock, so I kept looking carefully from one to the other. The two looked virtually identical. I knew that the leaves of sweet cicely are a shade of green lighter, and they develop faint white spots of mildew as the season goes on, whereas hemlock stays true green, but at the time of year we were looking for it, the slight difference in tone was the only way to tell them apart. It was the smell that clinched it, but Mum had already been munching on it, and she gave me some to taste.

As she has never ceased to remind me ever since, it was all Mum's doing; she gets just as enthusiastic as me and loves the hunt for a new plant too. It turned out that she had been on the same quest in the Ryburn Valley, near her own house, which is where we pick it now.

Sweet Cicely

I knew immediately, just as Mum had said, that we'd found the right plant, because sweet cicely really lives up to the first part of its name. In wartime, when sugar was in short supply, it was used instead. It can help to take the edge off rhubarb or gooseberries too; I now treat myself regularly to sweet cicely and rhubarb crumble. It tastes strongly of aniseed, but a lot of that gets lost when you cook with it, so you're just left with the sweetness – and it is ridiculously sweet.

The slightly aniseed flavour of sweet cicely also makes it go really well with fennel. When you combine the savoury aniseed taste of the fennel and the sweeter aniseed notes from the sweet cicely, they make a fantastic chutney. It is also a folk remedy for asthma and other breathing complaints, and you can use it as a sugar substitute for diabetics. There's no end to the power of plants.

Locating and identifying your prize is only the start of the process. I've then got to work out how best to preserve it, and how to extract the maximum flavour. There are no hard and fast rules, because every plant is different. For example, sweet cicely will last forever in my fridge, but is absolutely rubbish if you infuse it in water. I've had a gallon stockpot stuffed full of it and still couldn't squeeze out any flavour. The reason it lasts so long in the fridge is that it's resistant to water. Its molecular structure doesn't break down in it, so the taste can only be released by oil or alcohol – which is why it's so fabulous in gin.

Chefs like to understand all of this, because how to store, keep, process and preserve ingredients is all part of the same equation for them.

Dock

Everyone who's found themselves on the wrong end of a nettle knows to reach for a dock leaf. Mum and Nana did when my brother and I were sitting on the well outside Nana and Grandad's house which just happened to be sandwiched in the middle of a host of nettles. For some reason – and I remember it very clearly – a horde of what we called 'bloodsuckers' came flying towards us. I now know them to be common red soldier beetles – totally harmless, and veggies not carnivores – but at the time, terrified that we were going to be bled white and eaten alive by them, despite only wearing shorts and T-shirts it seemed a far better option to dive into the five-foot-high stinging nettles. We then screamed so loud that the noise penetrated not only the thick stone walls of the cottage but the din of the Saturday afternoon wrestling at full blast on the telly. Mum and Nana came running out to see what was the matter, and it was dock leaf action stations. Instead of just rubbing them on us, they spat on them too for extra strength. After that it was out with the emulsion paint brushes and we were covered from head to toe in calamine lotion.

I am mainly immune to nettle stings these days, which is a blessed relief, but since the dawn of time, folk have known that dock leaves soothe nettle stings, without knowing why. More often than not, though, traditional remedies turn out to have a solid scientific basis. It turns out that docks contain an antihistamine, and the agitation of the leaf is what releases the miracle cure.

You can also eat dock leaves, and dock pudding and bistort pudding (another kind of dock) have really strong cultural roots in West Yorkshire. In Huddersfield they even have an annual bistort and dock pudding competition. They have a flavour and texture a bit like cooked spinach, and used to be quite widely eaten in the north. Combined with nettles, oatmeal, onions, butter and seasoning, dock or bistort pudding still makes a tasty addition to a cooked breakfast.

I've made dock pudding myself, and until they've been prepared and cooked, the leaves are the most bitter and astringent things you could find, making me think that people must have been really hungry to have persevered until they found a way to make them edible – but I guess a lot of people were that hungry back in the day. A lot of chefs don't bother with them, but Adam Reid at the French in Manchester asked me to find him some and conjured up some heavenly little bistort and dock dumplings with them. However, they took a lot of finding, as although there were plenty about I discovered that a lot suffer from rust – a form of blight – which isn't very appealing

Back in the day when there were no fridges to preserve cheese, butter and the like, it was quite common to wrap store cupboard goods in large leaves – they still wrap Cornish Yarg cheese in nettle leaves to this day. Similarly, the leaves of the broad-leaved dock were once used to wrap butter, hence one of its common names – butter dock.

Wormwood and Mugwort

The name wormwood tells you its own story. Having made it into an infusion to get rid of intestinal worms, people then discovered that they liked the taste so much that they began experimenting with it as an ingredient in drinks. Among them – as its Latin name, *Artemisia absinthium*, suggests – was absinthe, and what we now call vermouth, which combines the German words for worm and wood. So not only did it get rid of unwelcome guests in your stomach, it made a cracking Negroni, but it contains mind-altering compounds as well, including a toxic substance called thujone. As a result, although Toulouse-Lautrec couldn't get enough of it, absinthe was banned in the US and much of Europe for many years, though the ban has now been rescinded, probably because they've forgotten about what it does.

Infused wormwood had such a nice flavour that people began mixing it with botanicals like apple, cinnamon and all sorts of other stuff, and stumbled on the recipes for a whole host of seductive Italian aperitifs like Martini, Campari and Aperol. The latter was first made by a pair of Italian brothers in 1919, but Martini and Campari are based on really old infusions. Although the modern versions tend to be quite sweet, if you taste the original recipes – and Campari kindly sent me some bottles of their recreated 1757 recipe a little while ago – you'll find they are far drier and full of herby flavours. So originally Campari was just a drink made by some Italian guy with a bucket of wormwood and a few random botanicals to spare.

One of wormwood's cousins is mugwort (and yes, they do all sound like they belong in Harry Potter's medicine cabinet), an aromatic that's a bit like a cross between sage and something you've never heard of. It can grow to a few feet tall and is a very easy thing to pick when you're out and about. It works well as an alcoholic infusion or as a chopped herb.

I am quite a fan of mugwort, but I suspect that its name – and indeed the name wormwood – doesn't have quite the appeal that chefs are looking for on their menus, despite its aromatic qualities. It's all in the name sometimes, not in the flavour. Having said that, a chef friend of mine, Oli Martin at Hipping Hall, who is really big on foraging, featured mugwort in a dish when he was a runner-up in *MasterChef: The Professionals*. He used a marinade of mugwort and rapeseed oil for his starter of kelp-cured langoustine with baby plum tomatoes, topped with crispy reindeer moss, more mugwort and a langoustine bisque which had the three judges quite ecstatic.

Sweet Woodruff and Meadowsweet

When you're sourcing wild plants, their flavours and aromas can sometimes lull you into a false sense of security, because they are not necessarily a reliable guide to whether they're safe to use. So never mind what it looks like, smells like and tastes like, you have to know what compounds it contains. When you know that, then you know if it's safe for anyone to eat and if so, whether it's safe for everyone, such as people with allergies or food intolerances. Strawberries, for instance, taste gorgeous, but some people are mildly allergic to them. Other plants contain compounds that can be life-threatening.

Sweet woodruff is a woodland plant that does exactly what its name suggests: it's sweet, it grows in the woods, in rough ground. It's a very clever plant. It reacts to sunlight and the warmth of the sun beating down on its leaves makes it emit a heavenly vanilla scent, but it doesn't want that. So it waits till the tree canopy overhead opens its leaves to protect it from direct sunlight, which is why sweet woodruff only thrives in woodland. In the warmth and the shade it grows to about a foot high in some places, with lovely little white flowers in spring. It grows quite low to the ground, with rhizomes on its root system that extend right through the woods, which you can easily damage if you pick it carelessly. It needs to be cut, rather than just pulled up; if you do that, it won't grow back.

Its scent is fabulous, and back in the day they called it ladies' bedstraw because people slept on it rather than eating it, so they'd wake up in the morning with a beautiful smell in their nostrils, rather than the less beautiful odour that tends to pervade the surrounding atmosphere when you haven't had a bath in months. Whatever, it's a waste just to lie on it, because sweet woodruff is a wonderful wild plant with a heaven-sent aroma and taste – a heady combination of vanilla and tonka bean – and chefs love to have it in their repertoire.

Sweet Woodruff

The Germans use sweet woodruff to flavour a wheat beer, Berliner Weisse, and it also makes great vinaigrettes. You can mix it with citrus flavours and it gives a nice balance, taking the sour edge off them, so a lot of chefs use it now, and it has become a bit of a signature ingredient for me.

When we pick sweet woodruff, it smells of nothing at first. It's only as it dries that the beautiful aroma begins to make its presence felt. Dehydrating sweet woodruff is essential for it to release that magical flavour and fragrance. It's always best if it can be sun-dried, but chefs can also dry it under the hot lamps on 'the pass' in a restaurant. It needs to be tossed around to make sure it's dried evenly, then crumbled by hand or in the food mixer. But it takes about five to six kilos of fresh sweet woodruff to make one kilo of dried.

When you come across sweet woodruff in the woods, it is always untouched and you might wonder why animals don't touch it when they ravage everything else around it. This is because sweet woodruff contains a compound called coumarin. Wild animals have learned to steer well clear of plants containing high levels of coumarin; if they didn't, it would thin their blood to the point where any form of cut would cause them to bleed out and die. Similarly, it can be dangerous if not treated with care, or given to people with allergies or medical conditions, because coumarin is the basis for warfarin, used both as a rat poison and to thin the blood of patients at risk of clots and strokes.

Bison grass, which is used to flavour vodka, tonka beans, melilot (sweet clover) and meadowsweet also contain coumarin in varying quantities, and common hay does too.

Some chefs are real fans of hay-baking, but the results are quite safe because if they're cooking lamb, say, they'll wrap it in the hay for two or three days first and the moisture from the meat will break down the water-soluble coumarin. It is then further eliminated by the steam given off as the meat's being cooked, so by the time it reaches the plate, all you're left with is the beautiful taste of the hay-infused meat.

As summer advances, the hedgerows, ditches and damp corners of fields are suddenly full of the frothy, cream-coloured drifts of meadowsweet flowers. The plant lives up to its name with that heady, incredibly sweet scent carried on the breeze – one of the signature notes of a British summer. A drink that Germans call *Maywein* – May wine to us – as the name suggests, was traditionally drunk on May Day to greet the start of summer. It was originally made from still and sparkling Riesling wine infused with meadowsweet, although it's now more commonly made with an infusion of sweet woodruff, known as *Waldmeister* in Germany.

When they added meadowsweet, they not only got a delicious drink, but found they could drink quite a few glasses and still wake up the next morning without a hangover. They probably couldn't work out why, but it's because as well as containing coumarin, like the bark of the willow, meadowsweet contains salicylic acid, the active ingredient for aspirin. There are quite a few members of that plant family that will sort out your headache, so perhaps we should spend less time heading to the chemist for the synthetic version when a wine and meadowsweet cocktail might be just as effective, and much more fun.

Mead wine (which means 'from the meadow') is now usually just made from honey but originally contained meadowsweet as well, and was traditionally used as a honeymoon drink. A honeymoon is what it says on the tin: get married when there is a full moon and drink alcohol made from meadowsweet and honey. Another popularity of mead wine back in the day was that since the coumarin in the meadowsweet thinned the blood and aided blood flow and the aspirin content got rid of headaches, there were absolutely no excuses for either party not living up to expectations on the wedding night.

Mead's still very popular down in Cornwall, and they have 'meaderies' where you basically go and eat chicken and chips with your fingers with big glasses of mead wine – which these days comes in all forms and flavours. When I lived in Pendeen, we had Trewellard Meadery down the road from us. The meaderies have their roots in mead wine history but these days tend to be more like medieval theme parks.

However, meadowsweet's effects can also be very damaging. It grows quite widely in America, and ranchers there used to dry it out as a winter cattle feed with medicinal properties, but if it goes mouldy, it's terribly dangerous. If I needed any reminder of that, the case of one Midwest farmer would provide it. He was de-horning his herd, and lost every single one of his cows because the hay that they'd been fed contained a large amount of meadowsweet. It had gone fusty and the mould reacted with the coumarin, increasing its content dramatically.

The drying process also concentrates the coumarin compound and it's really important that chefs understand that, so they don't use it too freely – about ten grams per litre or kilo is the safe amount. It's the same with any other plants and herbs I supply that could be dangerous in large quantities; without that knowledge, chefs might be putting their customers at risk. If, for example, they give too much to an elderly guest who's on blood-thinning drugs for heart or circulation problems, they are effectively increasing their dose, perhaps to damaging levels. A woman who is eight months pregnant with a history of miscarriages might be similarly at risk, and the consequences to both them and the restaurant where they have eaten are potentially catastrophic. It's just one example of the level of knowledge you require if you're going to start eating wild foods yourself or supplying them to others.

However, used correctly, and I stress correctly, meadow-sweet, bison grass, tonka beans, melilot and sweet woodruff are no more dangerous than any other kitchen ingredient.

Chicken of the Woods

In June the beautiful brackets of the chicken of the woods will start appearing on the trunks of trees. They are a ridiculous colour, almost unnatural-looking: bright, sulphurous yellow on the bottom and egg-yolk orange on the top. They grow like a folded omelette and I have seen them grow to about fifty kilos in weight. It's a pretty impressive sight and I had to take a picture to prove it.

If you find a lovely young specimen of chicken of the woods, the best thing to look out for is that the mushroom is plump and not dried out on the ends. All bracket fungi have a tendency to get rock hard, and the dry ends will be tough and not good to use. When you cut it from the tree, if the juices from the mushroom run down your arm then you know you have a good specimen.

Chicken of the woods are flavoursome and very versatile mushrooms but they have to be processed before they can be eaten. A woman phoned me up out of the blue a little while ago. I didn't know her; she'd tracked me down online. She'd eaten chicken of the woods in a restaurant and been violently ill afterwards, and I think she was wondering if she could sue the chef. I said to her, 'The thing is, you have to poach it in milk, not water, because there's something about the proteins in milk – probably its alkalinity – that draws any toxins out of the mushroom. So you need to ask the restaurant how they prepared and cooked their chicken of the woods. If they're not doing it right, they need to be told before anyone else gets ill.' Cooked correctly it's a really delicious mushroom, one of the most popular to be found anywhere because of its great flavour and texture, and yes, very much chickeny! Makes a good home-made schnitzel too.

Chicken of the Woods

One word of caution, though. It's vitally important not just to be able to recognise and identify the mushrooms correctly, but also the trees they are growing on. If a chicken of the woods has been growing on a yew tree, it will be as poisonous as the yew itself. Surprisingly, this is not because it has taken on the taxines from inside the tree but because the yew needles fall off and get stuck in the fast-growing mushroom, so you could cook it and not know the needle was in there. This is why we pick everything ourselves. It's not just what you're picking, it's where it's come from, what it's grown on and what's been picked with it or alongside it. Mistakes can be ruinous.

Oyster Mushrooms

We pick a lot of oyster mushrooms, which always grow on dead and decaying wood in the wild, but you can cultivate them as well. A lot of people do; they will thrive on anything containing cellulose, not just wood, but straw or even cardboard as well – at last, a use for those empty loo roll tubes.

There are a few oyster mushroom logs in my local woods but you have to be quick; sometimes I've known an entire log to disappear – oyster mushrooms, rotting wood, the lot. At other times I'll just be getting out of the car to go and pick some when I see a very smug-looking lady walking back to her car loaded with them – it's log wars out there sometimes. When I get there first, I'm the one with the smug look on the way back to the car park. We exchange smiles and pleasant looks, but she knows that I know and I know that she knows. That's how it is in the mushroom world sometimes.

The irony is I don't sell oyster mushrooms because you can get them cultivated, but I still pick them for my own use. There are two types. The one I usually pick is cornucopia, a branched variety with a number of stalks emerging from the main one. The other is the more familiar ostreatus, with the caps directly attached to the central stalk.

Milk-cap Mushrooms

Woolly milk-cap mushrooms, also known as bearded milk-caps, are prolific where I live. They seep a white fluid, a natural latex, when they are cut. You tend to find them under birch trees and when you pick them up, the underside is all fluffy. But don't be taken in by the fact that they look like the world's only cute and cuddly mushroom – the first one I tasted practically burned my tongue off. I can't help myself, it's what I do. I'd seen the local Polish ladies collecting them by the bagful so I was curious, until I found out you have to boil the life out of them for hours, making them not worth the hassle. The irritant they contain – velleral, a kind of dialdehyde thought to have evolved as a deterrent to hungry animals – is so ferocious that it can blister your tongue. Their Latin name – *Lactarius torminosus* – should have given me a bit of a clue, not to mention the early English one: 'bellyache milk-stool'. Yet they're prized and enjoyed for their peppery taste throughout Finland, Scandinavia and Russia, where they preserve them and eliminate the velleral toxin either by boiling, or brining or pickling them first.

You can find its cousin, the saffron milk-cap, elsewhere in this country, usually under pine trees, though infuriatingly they don't seem to grow around us. They're the colour of carrots, and their Latin tag – *Lactarius deliciosus* – tells you how good they are to eat. Spanish, Catalan and Provençal cooks recommend that you fry them in plenty of butter and then serve them with cream and pepper. They're not wrong – they're delicious, even if you can feel your arteries furring up with every mouthful.

Girolles

The demand from chefs for the beautiful, egg-yolk yellow and highly expensive girolles, also known as chanterelles, is almost insatiable. They're not only delicious but highly versatile too. Paul Leonard served them with cod loin, mussels and nasturtium at the Burlington in the Devonshire Arms at Bolton Abbey, and I'm sure he will be doing the same now he has moved to Forest Side in the Lake District.

Girolles

Girolles can now sometimes be found in high-end supermarkets like Waitrose, but we gather them in the wild and my best source of supply is not my native Yorkshire but north of the border. Like a lot of my best finds, I stumbled on them entirely by chance. Chris and I were staying at a posh hotel in Scotland for a friend's wedding one August. The ceremony wasn't until four o'clock in the afternoon, but I'd already slipped into my sequinned top and my best pair of trousers, and I had a bit of time to kill, so I was just standing in the bar, as one does, looking out of the window at the beautiful panoramic view. There was a wood in the middle distance, a really good mix of deciduous trees and pines – in this case Scots pines – maybe a mile or a mile and a half away, and I found myself thinking, 'That's got to be a good mushroom wood.' So I went to the reception desk and asked how to get to it, then borrowed a pair of wellies – they had a range of old wellies and boots they loaned to guests – and set off across the glen.

As I got nearer I could see that the dry-stone walls were smothered in moss, which is always a good sign because like moss, mushrooms thrive in cool and damp conditions. I passed some wood sorrel and got quite excited about that, because it normally sprouts in scattered patches and clumps, so you're on your hands and knees scratching around for it, but this was growing so thickly I could have taken a Flymo to it. But I gave myself a good shake – I was on the hunt for mushrooms – and carried on. As I made my way into the wood, I saw loads of little yellow patches at the bottom of all the banks. 'They can't be,' I thought to myself. Then, 'Wow! They are!'

The whole wood was a mass of girolles. I filled a couple of bags with them but I could probably have stayed there picking them for a week without really making a dent; I'd never seen so many in one place before. I was even happier than the bride and groom; Chris and I were having a lovely weekend in Scotland and would now come away with a tidy profit.

When I got back to the hotel, Chris started to say, 'Hurry up, love, the bride will be here any minute!' but the start of the ceremony didn't seem quite so urgent when I showed him the girolles and told him where they were. In fact he said, 'In that case, I'll just have time to pop down there for an hour myself then, before the kick-off.'

He came back with another couple of bagfuls and the next morning, before we set off for home, we headed out to the wood once more. The only slight drawback was that my borrowed wellies were so old and decrepit that they fell apart as we were striding back, but the staff didn't seem too bothered.

Wild Strawberries

Wild strawberries like to grow beside roads and paths, where every passing creature – dogs, foxes, badgers, even the odd human – has probably peed and pooed on them. So the best – in fact the only – place to pick them is under overhanging brambles. Nothing and nobody likes getting tangled up in brambles, so if you lift a mat of them and find wild strawberries growing underneath, you can be pretty confident that they'll taste as good as they look.

Again, you have to be careful not to take more than the plant can cope with. That varies enormously from species to species, and wild strawberries are particularly vulnerable. If you pick a green strawberry off a plant, you will devastate it. You may pick the third green strawberry off a stem and get away without killing the thing, but it really can be as particular as that.

The wild strawberry has another trick up its sleeve. I once got very excited when I found a huge patch of them, and made a mental note to revisit the place in a week or two to gather their fruit. I didn't realise at that point that there are two varieties of what look like strawberry plants growing in the wild. One is the absolutely gorgeous wild strawberry; the other is called the barren strawberry and – yes, the clue is yet again in the name – they are a complete waste of time. The leaves look the same – there is a slight difference in their structure, but it's a very subtle one – and the flowers look the same, but in fact they don't even belong to the same family, and they will never ever fruit. So you may stumble across what looks like a really wonderful patch in a south-facing slope with loads of sunshine to ripen the fruit, but hell will freeze over before you get a single berry from it.

*Wild
Strawberry*

The genuine articles are very fiddly to pick, and each one is about the size of your little fingernail, so you need an awful lot to make a bowlful – a patch of them would probably not yield more than two hundred grams of fruit at the most. It's worth the effort, though, because their flavour is exquisite – the concentrated essence of the best strawberry you ever ate, crossed with a hint of bubble gum. What's not to like?

There is nothing to compare with that taste of a wild strawberry, and that includes the shiny ones that supermarkets import from Spain and even further afield twelve months a year, which look great, cost a fortune and taste of absolutely nothing. Breeders have tried to replicate the taste of wild ones, and the Gariguette, an old French variety, probably leads the field. They come in trays, upside down, so all the points are standing proud. They're really expensive but they still don't taste as good as the ones we get our hands on if we're lucky enough to beat the birds and rabbits to the draw.

Wild Irish Strawberry Tree

You can find the fruit of the wild Irish strawberry tree in a couple of places around where we live. It can grow fifty feet tall and isn't actually a strawberry plant, of course. It's not native to Britain, but is naturally found around the Mediterranean, the Atlantic coasts of Portugal, Spain and France, and the south-western and north-western coasts of Ireland – where it's known as the Killarney tree. It can tolerate salt spray and likes a warm, temperate and humid climate, which is why it has continued to thrive even in northern climes where the Gulf Stream washes the shore and keeps the winter at bay.

Wealthy Victorian industrialists and the English landed gentry often imported unusual and exotic ornamental trees to decorate their gardens, arboretums and country estates, and Irish strawberry trees have been growing in many of them for well over a century. Fine specimens still flourish in a couple of our local stately homes – their microclimate must be warm enough to enable them to survive the Yorkshire chill. On my last visit to the nearest garden, one of the trees had white strawberries growing on it, and the other had red.

The flowers of the Irish strawberry tree take a whole year to mature into fruits, which ripen as the new flowers are appearing, in the autumn rather than the summer, and look like a cross between a strawberry and a lychee. Unlike an actual strawberry, they are quite crisp on the outside, with yellow-orange flesh. They taste like a blend of strawberry and guava, and are very high in sugars and antioxidants like vitamin C and beta-carotene. In Mediterranean countries where they do grow wild they're mainly used to flavour yoghurt or make jams and alcohol. The Portuguese concoct a potent aquavit called *aguardente de medronhos*, which roughly translates as 'strawberry firewater' and gives you a fair idea of what to expect if you get some down your neck.

They're not heavy croppers at the best of times, and even less so when growing in Britain at the northern end of their range, so I can never pick more than two or three kilos of their fruits altogether. That rarity makes them fit for a special occasion. If one of my chefs wants to produce something unusual with a bit of a wow factor, he could do a lot worse.

*

Bilberries

Bilberries – also known as blaeberries, whortleberries, whin-berries and a whole load of other names – are fabulous in all sorts of desserts and have become a bit of a family obsession. Although Grandad can no longer come with us, forty years on from the days when I was a toddler picking bilberries on Norland Moor with him, I still go up on the tops with Mum. Give any member of my family a bilberry scooper and send them on a walk on the moors and, without even really being aware that they've been picking them, they will come back with a bag full of bilberries. It has even been passed on to the fourth generation too, as my nephew Ryan can't go past a bilberry bush without getting excited at the sight of the big juicy berries just waiting to be picked.

After a hard day's picking on the moor, just before the sun sets, Mum and I like to sit on the bench above the Lad-stone Rock and just take in the beauty of the Ryburn Valley, with our dogs at our feet, tired out by having been out in the fresh air all day. They are always happiest being out and about and if they get bored while we're sitting there, there is a vast expanse of moor to explore, so they can trundle off for a few more minutes, but when you turn around, there is your dog flaked out in the heather again with a big smile on his face.

If rain clouds are approaching, we can spot them coming up from the valley and can get down the rocks and back to the car in five minutes, just in time before the first drops start to fall. Otherwise we sit and chat for ages, with our nets full of bilberries – and a few bugs. There is something especially satisfying about that moment together before we make our way home for tea. It seems to make everything worthwhile. A good night's sleep always follows one of those big days out in the fresh air; like I always say to Mum, 'You can't have a sleepless night, if you've run a marathon before bedtime.'

The climate on the moors above Norland is sub-arctic, and on winter days with what we call a 'lazy' east wind blowing – the one that can't be bothered to go around you and just cuts straight through you instead – you're left in no doubt why it's called that. Apart from the temperature, you can tell when you reach that sub-arctic level, because there are certain things that only grow there, like cotton grass. Literally thousands of silky white flower heads blanket any wet and marshy ground, but drop a hundred feet and there won't be a single one. They simply will not grow unless they're in the wet, cold and often icy conditions that would kill most other plants.

Bilberries thrive there, but will grow at lower levels too, though it makes me laugh when you see 'experts' on gardening programmes saying, 'Get some bilberry plants in your garden.' They will grow there and they'll probably be nice and bushy, but they'll never produce a single berry; they have to be in a sub-arctic climate before they'll flower and fruit.

On the top of the moor there used to be a set of ornate iron railings surrounding a concrete platform with a manhole cover set into it. When I was young, we were always up there, because it's where the best berries seemed to grow. One day the manhole cover was missing and the shaft beneath it was so deep that I couldn't even see the bottom. I dropped a stone and it was falling for ages before there was a very faint splash. When I realised how deep it was, I started looking at the crumbling concrete around the cover, and the sparsely grassed ground just beyond it, and said to my mum, 'I think we need to get off this straight away. I bet there's only a thin covering on top of it. We could fall through here, and it's a long way down if we do.'

We got off it sharpish and didn't go back for quite some time. It turned out to be the air shaft of a bunker left over from the Second World War. There was another entrance to it, protected by an iron gate, near a copse of pine trees much further down the hillside. It opened on to a broad tunnel that went right under the moor to the underground vault where ammunition was stored – if it had ever gone off, there would have been bilberries in outer space. My mum already knew there was an entrance there, but had never made the connection between it and the manhole and shaft on the top of the moor. My brother actually ventured deep inside the tunnel when he was in his teens, groping his way through the pitch darkness with only the dim, yellowing beam from a torch with a nearly flat battery, while icy water dripped on to him from the roof. He found some rusty old iron bunk beds where the unfortunate guards must have had to get their heads down.

Lingonberry

We often find lingonberries, another sub-arctic fruit, grow-ing alongside bilberries up on the moor, though they ripen a bit later in the year. Depending on where you come from, they're also known as mountain cranberry or cowberry and they're abundant throughout Scandinavia, northern Russia and North America. Most people don't expect to find them in Yorkshire, perhaps because they tend to associate them with the IKEA food counter, served with Swedish meat-balls, mash and gravy, but they do grow quite widely on the heather moors here and probably owe them their name, since ling – from the Norse *lyngr* – is the Yorkshire for heather.

Lingonberry

Raw lingonberries can be a little tart and mouth-puckering, but they are something of a miracle fruit. Studies have even shown that they may inhibit certain cancers.

They are also so full of natural preservatives like pectin and benzoic acid that they can be stored simply by keeping them in a jar full of water. Ancient mariners from the Vikings onwards have taken them on sea voyages because the berries last so long and, mixed with sugar, the water that preserves them can make a jam that doesn't even need cooking. They're also full of lots of beneficial vitamins, including vitamin C, which meant those Viking sailors were also protected from scurvy.

These days they are most often found as a syrup, which goes well with game, red meat, and all sorts of desserts. In Scandinavia they are used to make Lillehammer berry liqueur, and in many East European countries, including Russia, you can drink lingonberry vodka or a cocktail of vodka and lingonberry juice called *mors*.

Wild Cherries

We have masses of wild red sour cherries and wild black cherries near us. You tend to find them popping out of ancient hedgerows, but we've also discovered them on old pit heaps that have been reclaimed and covered in topsoil. The red cherries are consistently sour in a good way, which makes them perfect for preserving, but the black ones cover the full flavour spectrum and it's what we call cherry roulette. If you've got two or three trees growing close together, with their branches and fruit intermingled, you might have one fabulous, one OK and one horrible, and you have to keep tasting them to be sure that what you're picking is not eye-wateringly bitter. Keeping an eye on the branches of the good cherry tree takes a lot of doing when all you have in front of you are several leafy cherry trees with intermingled branches. But chefs will soon tell you when you have picked a bitter one. A lot of people think wild cherries are poisonous, but they're not – it's just that some are so vile you have to scramble around trying to find a sweet one to take the horrible taste away.

The best way to pick cherries without damaging their branches is to run your hand carefully through them like a comb and pull down gently so they drop straight into your net. It's best to keep the stalks on; if you remove them, the fruit will deteriorate more quickly. I once thought I would have a go at making cocktail cherries for Christmas, so I recruited a chef to help me. I popped round to his house one day, as he was excited about them maturing. He'd had some brewing in a concoction for three months and figured they were ready. As I arrived, he was changing his two-year-old's nappy and without washing his hands, he went out into the garage, put his hand into a giant home-brew tub, pulled out a handful of these cocktail cherries and offered them to me. My face just sank, and even more so when he told me he had infused them with a special hot chilli mix. On the grounds of the chilli I said I wouldn't be able to try one, and I never went back.

Rowan

We pick a whole range of autumn berries, and though people think that if they taste a bit bitter you simply have to put a load of sugar with them, that isn't necessarily true. It does work with some berries, but not others; it all comes back to their cellular structure and how you process them.

Take the berries of the rowan, also known as mountain ash. The trees prefer cooler climates but are widespread throughout Britain, northern Europe, northern Asia and the upland areas of the Balkans. Their clusters of vivid red or orange berries look beautiful but they're full of white fluff inside. If that doesn't put you off, unlike a grape or a raspberry you can't just pop them into your mouth and scoff. That's partly because they taste horrible, incredibly bitter, if you try to eat them raw, and they're toxic as well – not enough to kill you, but enough to make you feel really poorly.

Rowan

So you have to process them first. The toxicity fades if you heat them, or, to a lesser extent, if you freeze them. The other method, the one – surprise! – that I prefer, is to soak them in alcohol for at least three days, which starts to break down the composition of the parasorbic acid they contain, converting it to the benign sorbic acid. It takes some doing, as after that you have to freeze them and then heat them.

You can then sugar the berries and use them to make game jelly, jam or whatever you want from them. Things like that always make me think, 'Who the hell went through all the trouble to discover that in the first place?' But it's really worth persevering with rowan berries because they're a bit of a superfood. In traditional medicine they were used to treat sore throats and chest complaints. They are rich in vitamin C and a cocktail of anti-oxidants that have anti-bacterial properties and are claimed to boost the immune system, improve skin condition, eyesight, cell regeneration and healing rates. Poles, Czechs and Germans also use them to make schnapps or vodka, which is definitely my kind of medicine.

Hawthorn

We pick the berries of the hawthorn in autumn and the young leaf shoots in spring. The berries – which are always known as haws – are absolutely delicious, with a flavour like tiny apples. I hate picking them though, because you can't just run your finger down the stem like you would with a lot of other fruits. Hawthorn branches are full of vicious spikes, as hard as steel, so you have to gather them extremely carefully, one by one.

Once you have staunched the blood-flow from your wounds – note to self, keep up-to-date with the tetanus jabs – you can use the haws to make a compote, wine, ketchup or game jelly, but their flesh is like cotton wool and while you can cook them as if you were making jam, you then have to let the resulting purée drip through muslin over the course of twenty-four hours, to make sure you end up with a clear jelly and not a cloudy one. Haws don't give a monkey's about the winter chill, either, but I don't pick them when they're black – that's a step too far. I can still find places where they are protected from the worst of the weather, and a dull red. That's when you can tell they've been bletted and the frosts have frozen the berries, breaking down the sugars and carbohydrates inside them and intensifying the sweetness.

The pectin levels in haws are quite high, though not as high as in crab apples. There are over twenty varieties of those and we pick miniature varieties called ornamentals as well. Normal apples oxidise and go brown very soon after you expose the white flesh to air, but most crab apples don't, no matter how long after you've cut into them. They have the highest pectin level of any fruit; if you want something to set, add some crab apple to the mix. I learned that the hard way, because I once made three hundred bottles of crab apple syrup and then discovered they'd all set so solid I had in effect made three hundred bottles of jam. But when you're making hawthorn berry game jelly, if you add a little crab apple you'll not only have a really nice combination, you can also be sure it will set.

Wild Plums

We have wild plums all around us in Doncaster, both purple and yellow. The French call the latter *mirabelles*, but being from Yorkshire, we just call them yellow plums. When you're picking them yourself, the colour is really vivid compared to how they usually look in the shops – but none of us would look our sparkling best after being picked, packed and shipped in and out of a warehouse, before reaching the shelves. It's a bit like what happens with red potatoes; if you dig them up yourself, they are an intense pink, but after being shipped and stored for a few days they take on a dull, earthy look instead. Damsons are the same. We have a wild damson tree not far from where we live, and in the autumn sunshine, when the fruit is ripe, they look absolutely stunning – as if someone's started early on the Christmas decorations and hung rubies all over it.

Wild plums vary slightly in colour and size from tree to tree. Some are a little bit darker, some paler, some a bit orange; there are miniature ones as well, and they often hybridise. You'll find them in hedgerows, and once you know what you're looking for, you'll be surprised at how many there are. They're bigger than sloes but they can look quite similar, so you'll sometimes hear people say, 'Wow, look at those huge sloes,' not realising what they are. They taste fabulous too. The skin is often slightly more bitter than a commercially grown plum but the flesh is sweeter and much more intensely flavoured. So they are nothing like the ones you tend to find in a supermarket, but then wild plum trees are nothing like the ones you buy from a garden centre either – and that's true of all the fruits we pick.

We often use a stiff-bristled yard brush to rattle the branches of wild plum trees and get them to give up their fruit, but my last words to Chris as he walked out of the back door to gather some one day, were, 'Don't you lose that new yard-brush. There are already two stuck in the branches up where we pick them and I've only just bought that one.'

It should only have taken him an hour or so to pick the plums, and when he'd been gone for more than two hours, I was a bit worried and called his mobile but got no reply. The reason did not become apparent until he finally got home another hour after that. He had gathered a lot of good plums, but the ones higher up the tree needed a good shoving with the brush to dislodge them and while doing so, the inevitable happened and he got the yard-brush stuck in the branches.

Knowing what I'd say if he came home having lost another brand new brush, he decided to climb up the tree and retrieve it. He'd climbed to within a few yards of it when he got his foot stuck in the fork of a branch and found he couldn't go up and couldn't go down either. For a man in his mid-fifties to be stuck up a tree was a bit embarrassing, but when a group of ten-year-olds rode by on their bikes, Chris had no option but to swallow his pride and call out to them. It took them a moment to spot him because he was so high up, he was hidden by the foliage.

'What are you doing up there, mister?' one of them said.

Chris explained and after a few chuckles at his plight, since ten-year-old boys generally need no invitation to climb trees, they volunteered to rescue him. So they climbed up, unwedged his foot, and helped him locate the branches and footholds on the way down. Once he was back on solid earth, one of them very kindly went back up the tree to retrieve the new brush. Chris was pretty embarrassed at having to be rescued by a bunch of kids, but it was evidently better than getting bollocked by me for losing the brush.

When I go up there now picking wild plums I can still see the other two yard-brushes wedged high in the branches, but even though they are perfectly good brushes, I've decided it's best to leave them there. I try to ignore them but it still bugs me no end.

Rosehips

We gather the hips from the wild dog roses and sweetbriars in the hedgerows and woodland fringes on the lower slopes of the valleys nearby. You find them everywhere, but they're particularly abundant in the ancient hedgerows surrounding Doncaster. Much of the land has been in the same families for generations, and most of the hedgerows have been left to their own devices for hundreds of years. So a multitude of species of plants, shrubs, trees and wildlife are thriving there.

Rosehips are a favourite winter food for birds because they're really high in natural sugars and full of vitamin C, and they're good for humans too; I absolutely love them. Back in the day, there was a lack of oranges, lemons and other vitamin C-packed fruit in winter – at least, at prices ordinary people could afford – so they used to make a syrup out of rosehips, which you could buy in chemists when I was young. It would have knocked scurvy on the head, and helped people get over winter coughs, colds and flu in the days before antibiotics. The key to pretty much any home-made medicine is using ingredients that are high in vitamins and local honey.

After the Second World War, because of the shortage of vitamin C, the government launched the Rose-Hip Collectors

Club nationwide. People all over the country picked rosehips and got paid '3d per pound', with schools acting as central collecting points. Nobody had forgotten the lessons of history, particularly the disease of scurvy, and with imported citrus fruits virtually impossible to come by, the humble rosehip had to supply children's vitamins. Virtually nobody remembers the club now; like a lot of common knowledge about our fabulous British countryside, the memory has now been almost entirely lost.

The hips of dog roses – apparently so called as a result of the ancient belief that they were a cure for the bite of a mad dog – have a really distinctive taste: sweet, with more than a hint of Turkish delight, which, once tried, you will always associate with them. According to legend, Turkish delight was invented for a sultan with a rather challenging personal life. He asked his head chef to create something fabulous enough to put his large and disgruntled harem in a better mood. The chef mixed rose petals from the royal gardens with sugar and cornflour and the result is said to have done the trick. I took some homemade rosehip Turkish delight with acorns and homemade rosehip cough syrup on to *James Martin's Saturday Morning* in November 2019, and it went down a storm.

Rosehips work well too if you pop them into the freezer and blett them for a day or so, then mash them into a pulp to extract the juice and combine them with a stock syrup and a bit of fresh orange. If you want them to last a while, add a couple of grams per litre of potassium sorbate as a mould inhibitor and the same amount of citric acid as a preservative. I've had some last at least two years in a bottle. You can tailor-make your own medicine out of it too by adding eucalyptus or cloves; it's probably tastier than the stuff you can get from the chemist.

As a child my mum was a great believer in the power of disgusting medicine. I always ended up with the nasty brown stuff that tasted like a mixture of Marmite and Victory Vs and burned your throat on the way down. While my friends at school got the honey and lemon I got the nasty medicine because that way she could tell whether I was really poorly or just faking it to skive off school for the day.

Although the hips of dog roses begin to appear in high summer, you have to wait until they've been exposed to the first frosts of autumn before picking them, because that bletting brings out their sweetness. You pay a high price in the process, since they're surrounded by briars. You have to sort of weave your hands in and out of them in a usually unsuccessful attempt to avoid getting snagged on the thorns and bleeding all over them. It's a bit like playing the *Operation* game, but with razor wire.

I have an additional handicap when I'm getting stuck in: our failed truffle hound, Fred, can't get enough of rose-hips. More often than not, just as I've carefully eased my hands into the heart of the tangle of briars to bag a couple of particularly juicy hips, Fred will decide to help himself to a few low-hanging fruit, setting the briars twanging. No sooner have you freed one arm than you find you've snagged the other one in doing so, and meanwhile Fred will still be happily snaffling hips off the bottom branches. Must be something about dog fur because he never gets caught. He's daft, but not that daft.

There are other, very distinctively flavoured varieties of rosehip, and although they've all got thorns, most of them are nothing like as vicious as the dog roses. Unfortunately the burnet rose, usually found on coastal sand dunes or limestone pavements like the ones around some of the beauty spots of the Yorkshire Dales, is one of the exceptions. You don't get many hips on them but they're a really unusual dark purple, black or brown, and actually taste a bit chocolatey. You can make a tea, syrup or liqueur with them, but picking the hips should come with a hazard warning.

Japanese rosehips also grow wild in this country, despite their Far Eastern origins. They are vigorous enough to be used for hedging and are really resistant to pests and diseases. There are masses of them around us; they've completely overtaken the roadside verges in some parts. They have beautiful purple flowers and often grow alongside dog roses, but their hips have a completely different flavour. They are really big, the size of crab apples, look like squashed cherry tomatoes, and taste like them too.

Their thorns favour a different line of attack from the razor-wire – they leave you peppered with small, needle-sharp splinters, which are worse than a nuisance because if you don't get them out double quick, they'll infect your fingers at a rate of knots, and that's a bit of a worry because roses are notorious tetanus carriers. However, Japanese rosehips are well worth the trouble because the variety of dishes you can create from them is phenomenal. Chefs love them, though, because they're easy to process and really deliver on the fine-dining front. It's a real buzz to see something I have sweated blood and tears – literally – to gather, beautifully crafted on a plate with flavours that take you into a different world.

You start to get rosehips in August, but they'll last right through to January or February. They seem to be impervious to snow and ice, so they'll still be there, bright red and shiny, when most other things have rotted and fallen to the ground.

Sloes

The autumn hedgerows are full of the fruits of the blackthorn: sloe berries. It's tricky to pick sloes at the right time; you want to wait until they have a slight white bloom on them, like bilberries, and that's what tells you they're ripe and ready to be picked. You don't want to gather them before that because, although they look nice and purple and glossy, they're just not ready and will be as hard as nails.

It's quite easy to get sloes confused with bullace, a variety of wild plum that is nearly identical to a sloe, just a bit bigger. You think you have found the best sloes ever, but don't be fooled. Chris has fallen for this a few times and he's come back with bags of mixed sloes and bullace, which I inevitably have to spend a fair bit of time separating. A giveaway is that sloes are always on twig-like stems full of inch-long thorns – it's not called blackthorn for nothing.

By this point you might be thinking our fingers must be shredded from the gorse, juniper, hawthorn and rosehips, and you won't be surprised to find you're right. Although using gloves would be the sensible option, wearing them means you aren't very dextrous. And sometimes you just come across something that you need to pick there and then. It's just a case of suck it up.

One of the most popular uses for sloes is to flavour alcohol, but don't just stick to making gin, because vodka and white rum are equally good. You need to give it at least three months' brewing time, and remember that the higher the sugar content of the fruit you use, the lower the alcohol by volume (ABV) of the finished drink. So your forty per cent vodka can end up being about eighteen per cent ABV. Cut the sugar by half and get a better result: more of a fruit-flavoured liqueur with a kick than a sweet syrup.

Sloes

Most people know that if you're going to make sloe gin, you have to prick all the berries first to release the juice – but life's far too short to do it with the point of a needle, one at a time; give a whole load of them a good bashing with a brand new wire dog-brush. Just be careful to remember which brush is Fred's, and which is dog-hair free.

I have had some really good sloe jams and jellies made for me, and you can use sloe jelly as an alternative to cranberry sauce at Christmas. It's good with goose and game too.

The best way to make it is with the berries from your gin or vodka, as it won't take as long or be bitter because the alcohol the sloes have been steeped in for the last three months has done the hard work for you. Boil them up, add some sugar, a bit of lemon juice, and maybe a bit of crab apple or pectin to set it, and off you go.

Blackberries and Dewberries

Wild blackberries and the dewberries that hide among their thorny stems are favourite autumn fruits of ours. They start ripening at the end of August and can sometimes go right through to October. Dewberries are also called lucky berries, because finding one is a bit of a lottery. They're cultivated in America and can get to a really good size, but in the wild they're always quite small.

Loads of people go out picking blackberries in autumn, but almost everyone misses the dewberries beneath them. They always grow together – dewberries thrive in the same growing conditions as brambles, and mimic them to camouflage and protect themselves. People don't pick them because they look like a poor man's version of the blackberry: smaller and much less appetising. Don't be fooled, because they're bursting with flavour, a bit like a cross between a blackberry and a blueberry, and ten times as sweet. The only reason I don't gather them for my chefs is because they are so small and scattered you just can't get enough to make it worthwhile. When I pick dewberries, they're just for me.

When I'm out foraging, people are always interested in what I'm doing – also known as being really nosey – but I don't mind. In one of my favourite spots for gathering dew-berries, I often see an old lady who has a Jack Russell with a cowbell round his neck. I think he's a bit of a terror terrier and she put the bell on him to give everyone and everything fair warning that the dog is on his way. When he sets off running, the cowbell makes a right racket but she tells me he's a bit deaf anyway – well, if he wasn't before, he certainly would be now. She does power walking round the copse where I'm picking and usually stops for a chat with me after each circuit. She must be in her eighties but she says, 'Power walking keeps me fit.' I can vouch for that because she can fair shift. She often watches me picking dewberries for a few minutes and then says, 'Those berries you're picking are very inferior looking blackberries. Why don't you pick some of the nice juicy ones, like those over there?' For the life of her she can't work out why I keep filling my bowl with decrepit-looking berries but I've never enlightened her. I don't tell everybody everything, and dewberries are my secret and my treat.

Medlar

In the depths of winter, when almost all the other fruits have either been eaten or rotted away, one is still going strong. A distant relative of the quince and the crab apple, the medlar is without doubt the most minging of all fruits, so much so that the French call it *cul de chien* – dog's arse – for very obvious reasons. When you pick a medlar – it looks a bit like a persimmon (now more commonly known as a Sharon fruit) but with a more pronounced crown on top – you may think it's ripe and ready, but it will actually be rock hard and will take some time to rot down to the point when you can use it. There is no point going near them until they've rotted and are covered in mould and are in a squidgy mess at the bottom of a bucket or a garden sack. You have to wait until your medlars have been bletted, frozen and refrozen over and over again, and even after all that, it still take ages – weeks or even months – for them to deteriorate and to soften and ripen. We've tried to accelerate the process by sticking them in the freezer for a couple of weeks, and then putting them on a ledge above a radiator, and that has a good effect. Tradition is that a good indication is when fruit flies start hovering over them or a thick layer of mould appears, but I prefer to do without the mould and fruit flies, and get them just before.

Medlar

Medlar trees can grow to about twenty feet, but are often more like an overgrown bush. They are not particularly long-lived, no more than fifty years, and need warm summers and mild winters to thrive. They are only native to south-west Asia, Iran and the Black Sea coasts of Turkey and Bulgaria, but the ancient Greeks imported and grew them. They've been widely cultivated in Europe since then, and may have been brought to Britain by the Romans around two thousand years ago. I've heard claims that medlars have naturalised in some woods in the south-east of England, but I've yet to actually find one growing out in the wild. All the ones I know of are in the gardens and hedgerows of grand estates or very old houses. Quince trees and greengages do occur in the wild, but not round our parts, and I suspect that the three of them would have been the staple trees in the orchards of grand houses, abbeys and priories in the medieval era.

Back in the day they scraped the medlars off the ground when they fell off the trees in winter and then left them to rot, and I suppose the one saving grace was that at that time of year, the flies wouldn't have been around to lay their eggs on them. But there's a bit of a psychological barrier to overcome before you tuck in, because when they're ready to eat, their skin is wrinkled and the flesh looks really brown and manky, putrid even – just like a rotten apple that you would be putting straight in the bin. You can't imagine how anyone ever discovered that, but I guess if you're hungry enough you'll consume pretty much anything, and back in medieval times there'd be no shortage of starving people in the depths of winter willing to give them a go.

You can eat medlars raw or make jelly or medlar cheese – a paste which is not a million miles from lemon curd. They don't have any close equivalent in terms of flavour – a medlar is a medlar – but perhaps it's as close to a ripe date as anything. Aiden Byrne once did a pork terrine with a medlar cheese paste and it went down a storm. He sent me pictures all the way through the process; he was very pleased with the result and so was I. A few kilos of medlars goes a long way, and I think that will be on the menu for some time yet.

Sweet Chestnuts

Chestnuts from a shop, a market stall or even a street vendor with a brazier will most probably have been imported, either from France or, more likely these days, from China, and may well be rock hard, both because they're stale and because their shells are really, really thick.

Sweet Chestnut

I'm not suggesting that every locally sourced chestnut is a winner, though. A wood I use regularly offers a huge crop of magnificent and beautifully flavoured sweet chestnuts, while other trees in apparently similar locations have almost none at all. Those on south-facing slopes ripen quicker and yield bigger nuts, and if autumn gales are forecast during the picking season, I head for those in more exposed parts of the wood first, knowing that the ones in more sheltered spots can safely be left to last. My grandad always used to know where the best chestnut trees grew; I'd go there with him every autumn and come home with some right belters. A windy day in the wood can be quite hazardous though. The creaking of the branches as the wind tears through them, shaking off the chestnuts, can be quite disconcerting, and many a time we have narrowly avoided being hit by falling branches, but I'm of the mind-set that when it's your time, it's your time, so I'm not letting it put me off.

If you do venture into the woods to pick chestnuts, don't take your dog with you – the spikes on their outer casing are horrendous and it's murder on their feet. In fact, avoid woods for a few weeks in mid-autumn if at all possible if you have dogs. The tip for collecting chestnuts is to pick ones that look like they have freshly dropped to the ground in their green spiky cases. Gently put your boot on one and it will open naturally, the shiny deep brown nuts with tails on them appearing in a pair. Ease them out and pop them into your bag. You need to get them as soon as possible after they have fallen from the tree, because once they hit the ground there is an army of chestnut weevils just waiting for them to drop. They can't get in unless the nut has escaped its shell, and the tell-tale sign is a single pin-prick hole in the chestnut, so the trick is to check for holes. If they have got in, you won't know that you have them because it takes about three days for the weevil to munch its way through the chestnut and emerge as a grub. Great, your kitchen worktop is now home to a random grub, gross I know, but welcome to the joys of foraging. Also your bowl will now be full of what looks like sawdust – another giveaway.

Enjoy the chestnuts because the local ones have a sweetness in this country that is not found anywhere else.

Most people think they should roast fresh chestnuts as soon as they can lay their hands on them, but that's a mistake. Don't roast them, boil them instead, because then they don't explode in your oven and, even better, the shells become really pliable and you can just whip them straight off. I've shown some of my chefs how I prepare them, and they've been amazed at how easy to peel and how delicious they are. They make really good *marrons glacés* too, if you have the time. If you decide to peel them without boiling, put them in the fridge and use them within three days or, if you're saving them for your Christmas stuffing, freeze them as soon as possible. Chestnuts have tree nut pathogens so just use common sense and good hygiene practices and you will be fine.

The best way of storing chestnuts way into January is to layer them in newspapers in a cardboard box – so it's dark and warm but they can still breathe – and stick them in the shed. One of my autumn jobs is to clean up my foraging shed at the bottom of the garden, ready for winter projects. It's also great when I have friends over, and we often sit in the shed, lit by soft candlelight, with lanterns dangling from the beams as well. Outside there are hurricane lamps that look fab hanging from the washing line. I painted them, removed the wicks and put a penny in each one to act as a base for ever more candles. I boil up a bucket of chestnuts and we sit there on those dark, crisp, early winter evenings, peeling and eating chestnuts, usually accompanied by a large Negroni, while we put the world to rights.

Walnuts and Hazelnuts

As well as collecting fully formed walnuts in October, we also forage green walnuts in June. In Britain, we have two sorts of native walnut trees growing wild: English walnuts and French walnuts. We've also got black walnuts, but it's worth mentioning that they look like false walnuts. Walnuts originally came from Persia but are now naturalised throughout Europe. The Romans brought the trees to France, where they grow prolifically. There are four varieties – Marbot, Corne, Granjean and Franquette – in the Dordogne region alone, and many French towns hold annual *fêtes de noix* when they're ready for picking.

The French walnuts we have in England date all the way back to the aftermath of the Battle of Agincourt, over six hundred years ago. Once the battle had been won, as they made their way home, the English archers and soldiers had to feed themselves, so they pillaged the farms they passed, and foraged for whatever wild food they could find. It was October, the time of year when walnuts were ripe, so they tucked them into their knapsacks and lived off them when other food was scarce.

When they got back to their homes in England, many of

them planted their remaining walnuts. A lot wouldn't have taken, because you can't just plant a walnut in the ground and hope it will grow. You have to make like a squirrel first: agitate the nut, damage it, break the shell as if it's been half-eaten, and then you can plant it and you'll get a tree. If you take the nut completely out of its shell and plant it, that doesn't seem to work, and if you plant a whole, undamaged one, that won't grow either, but if you duff it up a bit first, that does seem to be the magic ingredient.

So lots of their walnuts must have failed to take root, but some must either have already been damaged or the English squirrels got to work on them. So now we still have our native English walnut trees, and French ones as well. You can tell them apart because the English have a thin shell and a large, rather almond-shaped nut, whereas the French have much thicker shells and a more 'boxy' shape.

I'm not just being partisan when I say the English variety are much nicer to eat. They're more defined, developed and sweeter, while the French walnuts here are very hit and miss. They must be fantastic in their native land because they devote whole festivals to them, but here they tend to have quite small nuts inside those massive armour-plated shells, or sometimes none at all. I suspect the reason is that the French trees evolved in a much gentler climate – perhaps even as direct descendants of the original Persian walnuts – and, despite having grown here through a dozen or so generations of mature trees, they are still struggling to cope with our colder, wetter conditions.

According to folklore, walnut trees thrive on being beaten, but there is no evidence that it's true, and the belief may derive from the fact that long poles were once used to dislodge the nuts from the tree and may also have been used to break longer branches in order to encourage the growth of shorter fruiting spurs. We still do a similar sort of thing, as our local council is obsessed with cutting off the low branches of trees just where the best berries and nuts are. It's probably driven by some health-and-safety-danger-to-passing-pedestrians-and-traffic nonsense, but it's a shame because it would be a really big help to communities like Doncaster to have easy access to otherwise expensive nuts and fruits that people could gather for free.

In early summer, after their flowers have been pollinated, the immature walnuts – what we call 'green walnuts' – begin to form. I think green walnuts are more versatile than fully-formed nuts. At this point, chefs can make green walnut ketchup with them. René Redzepi once asked me about it when we delivered a new crop of pink larch cones, and I told him you can process green hazelnuts for ketchup in exactly the same way. I gave him the recipe for it, but I don't know whether he actually made it.

Hazelnuts are sometimes called cobnuts or filberts, but they're all the same beast. At one time we used to wait until they'd ripened before trying to gather them. More fool us: they all dropped off the trees at the same time, and the rabbits and squirrels got there first and scoffed the lot.

Although I still pick some of the ripe nuts – since rabbits can't climb trees, if you get them while they're still on the branches that eliminates some of the competition and means you've only got the squirrels to worry about – after a few frustrating years, we decided to leave most of the ripe hazelnuts for the animals, and just pick some when they were still green. You can process these exactly the same way you do with green walnuts, so as well as ketchup, you can make wine and pickles. I don't know why people don't use them more often.

Green walnuts also make a fabulous wine, which is well worth waiting a few months before drinking it. I had Josh Overington of Le Cochon Aveugle in York lend me some to take down to *James Martin's Saturday Morning* and it really is like a top-end sherry. Or you can just pickle them in the traditional fashion – though I have to confess that whenever I've bought a jar at Christmas, it's lasted for at least five years, unlike cocktail cherries, where I can buy five jars and they'll all be gone before Santa arrives.

If you want to go the green walnut route, you need to move fast, because once the shell begins to take shape inside them, even though it's only very thin at first, it's absolutely rock hard. So you need to take a skewer and a pair of gloves with you, and pierce whatever you're planning to pick to make sure it's usable. If it slides through easily, you're in business, but if you hit a brick wall, you'll have to look elsewhere. The good news is that you might not have to move far, because of course the fruits on the south-facing side of a tree will ripen and mature faster than those on the north side.

'So why the gloves?' I hear you ask. Well, the flesh and the juice of the green walnut are both creamy white, but don't be fooled by that; if your hands are bare when you're picking and cutting them into quarters, they'll go the colour of brown furniture, and there's not much call for that stuff these days. And though the stain will gradually fade in about six weeks, nothing short of amputation will get rid of it in the meantime. I keep thinking that we should market the stuff as fake tan, because if you rubbed half a green walnut on your face you'd go the same colour as a mahogany cabinet.

Having harvested some green walnuts, we can return in the autumn to gather the mature nuts from the same trees. Grandad used to pick them for us every year, but most of the families I knew when I was young only seemed to have them at Christmas, mixed with a few hazelnuts and Brazil nuts in a bowl on the sideboard. I remember us kids watching the grown-ups wield the nutcrackers as they tried – and usually failed – to extract a whole one from its shell without shattering it into a thousand tiny pieces. Maybe that's why they were still there at Easter, coated in dust.

A favourite recipe of mine, pinched from Stephen Smith at Andrew Pern's Michelin-starred restaurant, The Star at Harome, is to boil my walnuts in a stock syrup for a bit, then deep fry them in clean oil and finally roll them in either sugar or salt. They are the most amazing walnuts you've ever eaten. As an even more indulgent treat, I like to roll some of my walnuts in caster sugar that I've infused with a whole sliced white honey truffle – it just takes them to another dimension.

Acorns

Most people would leave the fruits of the oak tree – acorns – for the squirrels, thinking that they're poisonous. They do contain a lot of tannins, which can make them taste bitter, but bitter isn't always poisonous. Acorns are perfectly palatable and quite versatile. They don't really taste of anything much when they're raw, other than being a bit astringent, but they are very nutritious and if I was desperate, I'd definitely eat them, though not too many in one go because, although I've never suffered from it, they can cause a bit of a stomach-ache. What people have forgotten is that once upon a time we ate acorns a lot. Then we became farmers and kept pigs and they ate acorns a lot. Now, not only do we not keep pigs, we don't eat acorns any more either. Funny really how the world goes round but comes to an abrupt stop, and then something perfectly edible gets the obligatory 'it's poisonous' tag on it.

Acorns fresh from the shell are bigger than peanuts, very dense and really light-coloured, yet if you put them in a pan and start to boil them, the tannins turn the water black as night in seconds. I've had to change it four or five times when I've been preparing them. The Germans made *ersatz* coffee from acorns in wartime when there was no proper coffee to be had, and you can immediately see why: apart from that jet-black colour, when roasted, they have a nutty, chicory taste, which would have been as close to the real thing as they could get. Once you've prepared the acorns – and I usually give up after five boilings, more out of boredom than com-pletely blitzing the tannins – they are no longer pale, almost white, but more of a deep, dark brown.

As well as grinding acorns up for coffee, you can make flour from them, a healthier alternative to wheat for coeliac sufferers. The wheat we know and use today isn't a naturally occurring crop and has only been around for about five hun-dred years. Before then, our ancestors made flatbreads like the Greek *pitta* breads that we are all now familiar with, or the Norwegian *flatbrød*.

In Victorian times, when they didn't have the vast acre-age of wheat fields we do now, they ate more cornbread, and used a variety of different roots and nuts, including acorns and chestnuts, to make flour. Chestnuts have a raising agent in them, and they're a complex carbohydrate, whereas an acorn's a protein. Chestnut flour is French and very expensive, but it's good and it's easy to make and use. Acorns, on the other hand, take an age to process, but if you've got the patience to persevere, it'll be well worth the trouble. I've not come across commercially produced acorn flour, so you're going to have to have a dabble yourself if you want to try it.

Acorns

It's a bit of a bind to do, though, as they need to be put into cold water with a change of water every other day or so. You can't boil them because that would change the structure of the nut and wouldn't make for good flour, so you need to do it the hard way. After about a month, though, you can peel the acorns and grind them. Then mix it with any other flour of your choice to bulk it out a bit and you have a really nutty flour.

I've often thought about going round with a leaf blower on suck so I can vacuum up all the acorns and take some of the hard work out of it, but I've not quite got around to it yet. Last year I sent young Ryan and my Mum acorning, as they have loads down their neck of the woods. While they are out, my mum sometimes lets Ryan get on with it and has a sit down for a while. She then rings me up and puts it on speakerphone so I can hear Ryan talking to himself while he's picking, saying, 'Oooh, look at the size of that one . . . Oooh, look up there at those . . . I'm going to have that one, it's a right big un.'

They have a competition to see who can get the biggest or the most. That's the thing about foraging, it brings generations together in pursuit of something enjoyable. As well as being on his Xbox, like most seventeen-year-olds, Ryan also enjoys the company of the different generations and being out with his nan picking acorns. Watching them trying to outdo each other is rather endearing.

Unfortunately my mum's dog, Bo, gets a bit bored by it all and when she catches sight of something, she is off, so Ryan sometimes has to down tools and hotfoot it into the distance to try and retrieve her, but that's what young legs are for.

Then it's a trek home past Nana and Grandad's. Grandad always wants to know exactly what they have picked and nods his head approvingly at the sight of eight kilos of acorns, knowing exactly what I'm going to be doing with them. Then we all have a sit down in the back garden with a can of pop and have another chat about foraging.

*

Silverweed

There are some plants that I would only pick if a chef was entering a cooking competition and wanted something a bit fancy pants to impress the judges. And it couldn't be any old competition, it would have to be a good one like National Chef of the Year or the Roux Scholarship, which can jet-propel a chef into the stratosphere. One such plant is silverweed, which is a very slender root that can be dried and made into flour.

Silverweed has really silvery foliage – again, the clue is in the name – and is so common that you often find it growing alongside paths. Folklore has it that footsore Roman soldiers used to line their sandals with it during the endless marches along those dead straight roads. It's one of the few plants that seems to flourish and spring back even when trampled, which may be how the legionaries realised it would make a nice cushion.

They used to pick it regularly in Victorian and earlier times, but they didn't have as many dogs as we do – and the chances are that the foxes and badgers who use the same paths while we're sleeping won't have been too shy to pee on it either – so while the Victorians probably just gave it a quick rinse and ate it anyway, I would only harvest it from areas that I know are animal-free. So far I have only picked it to see what the flavour is like – it tastes a bit like parsnip to me.

In the bad old days the peasants would roast, bake or boil the roots of silverweed to survive in times of famine. There weren't many other options; if they didn't have their own crop-bearing land, they had to depend on what they could forage. The rich ate meat, the poor chewed on silverweed. It wouldn't have tasted that great but it would have kept them alive. For most people in our affluent society, extracting flour from silverweed roots wouldn't be worth the effort, but it's worth knowing about just in case some future natural disaster leaves us scrabbling for whatever food we can find.

Chickweed

Chickweed is a plant that we can pick when the days grow shorter, because it's as common in Yorkshire as it is everywhere else. Otherwise known as winterweed, it flourishes on roadsides, coastal cliffs, riverbanks, gardens – you name it, chickweed grows there. It has a delicate salad leaf – a bit like watercress – with a really peppery bite, and it's bursting with vitamin C. Less appetisingly it's also a haven for slugs, but when everything else is wilting, shrivelling and expiring, in all but the most severe conditions chickweed just keeps on going. It's even been known to flower and set seed despite being covered with a layer of snow. It always amazes me how this very delicate little plant can survive, but it does.

It's a good thing that chickweed is everywhere, because that means it's not cut and cut again. Over the years I have learnt that if you cut a big patch of chickweed from one area before it's gone to seed, it won't grow back again, so you have to be really careful to leave a patch that can go to seed and regenerate itself. Otherwise your chickweed patches run out. Fortunately, when it's gone to seed it's prolific in regenerating itself, so it's fine as long as we tend our patches carefully. It's all about sustainability, after all, and it's always good to know these things.

Chefs love this winter salad plant, and it's an important one for those who rely heavily on foraged elements in their dishes. If they haven't prepared as well for winter continuity, chickweed is something that keeps the philosophy of wild food going throughout the year.

Reindeer Moss

Even in the apparently dead times of year there are still plants to gather. In November, for example, when most trees have shed their leaves and the vast majority of plants have withered and died back, lying dormant until the strengthening light and warming soil of the next spring coaxes them back to life again, many mosses thrive in the dark and damp conditions. We pick reindeer moss at that time of year – more properly called reindeer lichen because it isn't actually a moss – and it is beautiful. Like many lichens, reindeer moss is about six per cent acid, meaning you shouldn't eat it raw because it can cause stomach upsets, but if you either dehydrate it or fry it until it looks like glass and then season it, it's absolutely fabulous. I make sure I always supply my reindeer moss a bit grubby, because then the chefs *have* to wash it and prep it thoroughly. I'm not just a pretty face . . . In Scandinavia they also use it to make aquavit.

Reindeer moss is one of my mum's favourite things to eat and one of Chris's least favourite things to pick. So his face was a picture one day when I took a call from the head chef of a brand new restaurant, asking if we could supply him with forty kilos of reindeer moss. It is so light that it was like being asked to supply forty kilos of candy-floss – it would have filled a Luton van.

At such short notice I could only get a few kilos, and when I turned up to deliver them, I found myself in a beautifully decorated restaurant, with a large team of very busy chefs in the gleaming kitchen. They were what I call the tweezer brigade, and the more aspirational the restaurant, the bigger the tweezers seem to be. It was so loud in there, with music blasting out, that it was nigh on impossible to talk. I couldn't have worked in an environment as noisy as that – and I've worked on an aircraft carrier with Harrier jump jets taking off – but from what I could lip-read from the head chef, they seemed to be managing.

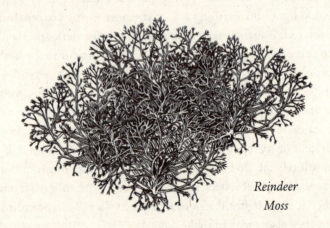

Reindeer
Moss

He was clearly determined that nothing should detract from the glamorous, uber-cool image he was trying to project, because when I later saw the menu he'd produced, the reindeer moss I'd supplied – Doncaster's finest – was listed on it as 'from the Isle of Skye'. Doncaster clearly didn't strike quite the right upmarket note . . .

When foraging you need to have footnotes in your brain about everything. It's always worth a note that there are two poisonous lichens and they are bright yellow, so the rule of thumb is don't eat bright yellow lichen. It's easier saying that than giving the Latin name.

The Conifer Family

Pines, firs and spruces are all members of the conifer family, and they're all edible, making them another potential source of food in winter.

That is, as long as you steer clear of the yew trees that look a little bit like them. I have a problem with chefs and yew. They are forever using yew platters to serve food on. There was a restaurant recently showing off fresh charcuterie served on a yew platter. All I could think was, 'Ye gods!' I make no apologies for preaching to chefs when it comes to safety. It's not all about what you pick sometimes, it's about how to prepare and how to serve it as well.

I've seen it before though. As I was doing the rounds a while ago, dropping off the orders at the restaurants I supply, I tootled into one Michelin-starred kitchen and stopped dead in my tracks as I saw a chef meticulously picking needles from the branch of a yew. So I said to the head chef, 'About that branch you're using.'

'You mean the Douglas fir?'

'Actually, it's not Douglas fir. It's yew and it's poisonous.'

'No, no,' he said, 'it's Douglas fir. I picked it myself, it smells of grapefruit.'

As he was a top chef I even doubted myself for a minute, so I went over and picked it up and took a really close look. 'Sorry,' I said, 'but it's yew.'

He still wasn't having it. 'I'll show you the tree,' he said, and took me outside and showed me the newly planted yew hedge he'd cut the branch from. I tried to explain in great detail why it was yew and not Douglas fir, but I still couldn't budge him and in the end I just had to give up. I don't know if he had second thoughts after I left, but as far as I know nobody died. I don't normally get involved, as nobody appreciates being told they're wrong, even if they are, but it was a bit of a moment where you really don't want to walk away and not say anything.

Yew contains toxins which can cause heart failure. It's not unknown for an animal to drop down dead after accidentally eating a part of a yew. Back in the day yew was used to make longbows. It had the flexibility of the inside wood and the rigidity of the outside wood, giving it tensile strength so it made a really effective weapon. Unfortunately though, the archers didn't last that long because as well as using the bows, they also tended to be involved in the process of making them. When you dry the wood out, you lose moisture, which intensifies the poisonous toxins; so the archers were inhaling poisonous dust.

To top it all off, yew is a softwood, so it's porous and if your charcuterie or whatever you're serving on the platter contains moisture, this could very easily leach out the taxines and make the food poisonous. So why anyone would put anything on yew as a serving platter is beyond me. It's not a dead cert, but why risk it?

Provided you manage to avoid yews, evergreen trees can be a lifesaver for people trapped in the wild. You might have been in an aircraft that had crashed into a snow-covered mountain in the dead of winter and left you with no hope of rescue, but if so, you don't have to eat each other, you could eat the pine trees first! The cones contain pine nuts, which are edible, of course, and when they're little they're very lemony and zesty, and you can eat the young cones as well, particularly of slow-growing species like the Cedar of Lebanon, which takes years for the cones to mature.

You can also eat the young green shoots of the new growth on the trees, and the needles of pine, spruce and fir trees at any time of year, or make a tea with an infusion of them – they all have a refreshing citrus taste and are a good source of vitamin C. Like the Christmas Tree Syrup I made for Lakeland, you can use the new growth from the trees to make a syrup that you can eat with roast grouse or duck, mix with tonic, soda or lemonade, or add a citrus zing to a Martini.

Remarkably, every single bit of a pine tree is edible. In Scandinavia, where the growing season is shorter and the climate more problematic for cereal crops, they even use the soft, resinous sapwood of pine trees to make flour. It consists of a white membrane called phloem just under the bark which acts as the tree's irrigation system, replenishing the pine needles with the plentiful water they need.

There are plenty of commercial saw mills in the country and pine is their staple diet. They strip off the bark first, so you can pop down, speak nicely to them, and help yourself to some phloem before it's been put through the chipper. If you dry it out, then grind it and mix it with another flour of your choice, you can make fantastic breads from it. Phloem contains a natural raising agent as well as a natural preservative; just mix it with water to make flatbread, and it'll keep well.

In desperate circumstances you could eat the phloem as it is; it won't taste particularly nice, but it will keep you alive. You can get sustenance from pine trees in even the most arctic conditions. If you eat snow to try and get water from it, the energy required to turn it into heat is a negative number, so you're expending more energy than you're gaining. However, pine needles contain a lot of water as well as oils, sugars and vitamins. They have a natural form of anti-freeze in them, so they're protected from frost, and the water and the nutrients they contain are immediately accessible. So they need a bit of bashing, but once you get past the tough outer casing of the needles that deters birds and animals from eating them, they make a lovely aromatic mush. So, like a cactus in the desert, pines can keep you alive in even the most extreme situation.

Juniper

Juniper is an evergreen, and since nothing much else is growing in the depths of winter, we often take the chance to go up to Scotland at that time of year and gather some. There are a few plants that deserve to be treated as special because they're so hard to get, so when you do find them, you make the most of them. For me, juniper is one of those special plants with a little bit of magic about it. We only ever go up once a year, in December, to pick it, so it's a bit of a Christmas treat for all

of us – and nothing says Christmas to me better than a traditional Scottish hotel with an all-inclusive bar. Apart from
juniper, Scotland is rich in wild foods and is a really special
place for foodies. There is a lot to be said for hunting and
gathering in a place with rivers full of salmon, seas with the
best shellfish in the world and hillsides full of deer. Not to
mention the oats and raspberries that grow there.

Juniper

When you're trying to find juniper, the best way to
describe what you're looking for is that the male tree is big
and phallic-looking, while the female is a short, squat, fat,
rambling bush that spreads all over the hillside. So you first
look for a plant that seems really out of place. It will be the
most perfect-looking conifer, tall, graceful and about ten

feet high, standing alone but surrounded by low, sprawling bushes. That's when you know you've found some juniper. Only the female plants have berries, so when you get close to it, you'll see that the tall, perfect conifer has no berries at all but the squat bushes around it will be covered in them.

Juniper is very slow-growing, taking seven years per foot, and needs to be quite high up to survive and thrive. We do have it in England – there's some in the Peak District – but I go all the way to Scotland rather than Derbyshire, which is much closer, because in the Peak District there's so little of it and it's so slow-growing that if I started picking it, the plants would never recover.

We once got asked to supply someone with tons of the stuff but I refused, even though I'd have made loads of money, because it would have been unsustainable. The plants wouldn't have regrown, so I'd have killed them off, and I'd rather do without the extra cash than have that on my conscience. If I was that way inclined, I could go out and ravage everything and people would probably be none the wiser. There's no audit, nobody is checking up on what I'm doing and there's usually nobody with me, but it's a matter of personal integrity; I couldn't live with myself if I acted like that, though I know of a few who do.

Although it's rare in England, the place where we go in Scotland has about four square miles of juniper. So we can take a bit from this bush and a bit from that, and make sure we rotate where we're picking from, so it all has the chance to recover. In fact, it does it good, because like most plants, juniper benefits from a trim and a bit of a thinning out.

The Scottish juniper has the best, most flavoursome berries. They look a bit like slightly larger peppercorns and taste, as you would imagine, like gin. When you put them under a magnifying glass, they are like miniature pine cones, tightly packed and full of a natural anti-freeze, so you can put them in cold storage and keep them in suspended animation until you need them. They don't actually freeze, but they don't deteriorate either; they're just suspended in time. That's really handy because it takes two years for a berry to mature.

You can use every part of the plant – the berries, needles, twigs, branches and even the trunk – because they all have exactly the same taste. Juniper is used to flavour gin, of course – it can't be called gin if it doesn't contain juniper. When you make gin, you utilise the volatility of the natural oils the berries release. To extract the maximum flavour you need to put them in a muslin bag and give them a good bashing before soaking them in a really high percentage alcohol.

Juniper can also be used in making whisky, but only as firewood. When they used to make moonshine whisky up on the Scottish moors and hillsides, they dodged the customs men by heating their stills over fires made with juniper wood because it made virtually no smoke as it burned.

Juniper is so rough and spiky that very few creatures have a taste for it. Highland cattle are a notable exception, and if they ate enough of it, their beef would have an even more amazing taste. I wonder if farmers have ever thought about charging a premium for it. However, because juniper grows quite densely on the hillsides where it thrives, the chance of running into Highland cattle lurking behind the bushes is fairly high and with a dodgy knee like mine, that's not a great idea.

My mum and I have been chased by many a herd of cattle over the years, and trying to shove an eight stone German shepherd over a wall before you can escape yourself, while the thunderous sound of hooves grows ever louder in your ears, is not an experience I would recommend to anyone of a nervous disposition. So my first step when entering a field is to look for big divots on the hillside caused by the weight of the cattle, and of course fresh cowpats. If they are crusty-looking, the flies and dung beetles have got to work on them and grass is starting to grow up through them, you're probably good to go, but if there are fresh ones it's not worth the risk in my book. You just have to know these things because it can be a matter of life or death. Don't tell me 'Oh, relax, they're just cows', because, according to the Health and Safety Executive, cows are the most dangerous large animals in Britain. Cows have killed an average of five people a year over the past fifteen years, compared with only two people a year killed by dogs.

Juniper bushes are densely populated by spiders that are so small you can barely see them. If I sat long enough and had a good think about it I could probably work out why. Perhaps it's simply because juniper is so spiky and inhospitable that birds don't gather on it.

When you leave juniper in your car, the spiders won't waste a second getting to work. You fire up the windscreen wipers because you think it's misty, and discover that it's their network of tiny webs covering the windscreen. The little buggers also infiltrate the foam in the roof, and keep getting in my hair as I drive along, so it's a good job they don't give me the heebie-jeebies.

After many years of foraging, there are unique ecosystems in each of our vehicles, as well as our house and our shed. My Citroen Berlingo is a haven for the juniper spiders, who have settled in nicely and are always very appreciative of any bugs escaping from our foraging bags. The small car we use for mushrooming is a retirement home for beetles, especially the small shiny green ones that look like the jewels in the Tetris computer game – I've never been bothered enough to look up their proper name. In the house we are plagued by daddy long legs, and the foraging shed at the bottom of the garden is full of giant spiders, and all sorts of other critters. However, it is all natural and no different from the environment in which we pick everything. I do tell my chefs that I cannot guarantee everything is bug-free, but we do our best.

The first time I went up north – and I mean way up, in serious Scotland: the Highlands, near Loch Ness – looking for juniper, I'd done a bit of research and a lot of legwork and discovered that it was particularly prolific around a place called Whitebridge. When I went there again with Chris and my mum, while I was busy meeting potential customers, they went out to gather the crop. I'd given them precise directions: 'When you get to Whitebridge, you've got to go up this path, through this gate and through a field, where there's an upturned picnic bench,' and so on. They did all that and even found the upturned picnic bench and everything, but they came back having failed to find a single sprig.

It turned out that there were two Whitebridges in Scotland, both of which boasted an upturned picnic bench in a field, so I'd sent them on a complete wild goose chase. The worst thing was that they were convinced they were on the right track because they'd found the bench and so they had been searching for nearly four hours, because they didn't want to come back empty-handed. I was back at the lodge when I realised what had gone wrong, but the phone signal is so rubbish in the Highlands that I couldn't tell them. They weren't best pleased when I finally got through to them, but I told them, 'Look on the bright side. Nothing wrong with a bit of extra exercise, and it has knackered the dogs, plus as compensation for your wasted day, I've just bought you a good bottle of single malt.' That always goes down well with my mum and Chris, who can both drink neat whisky like it's going out of fashion, whereas I'm more of a wine and sherry person. So it wasn't long before the dogs were fed and flaked out and Chris and Mum were sat in front of a big roaring fire, with a large whisky in their hands while they enjoyed their favourite pastime: berating me for getting things wrong.

The next day we went up the road in the opposite direction and found plenty of juniper growing on the side of a hill – just like that.

When we first went up to Scotland we would try to talk to the locals about juniper but none of them seemed to have a clue about it, even though they were surrounded by it. I thought that was a bit sad really. Those people lived in an area which was full of a really rare and special plant, and yet, even though a lot of places had names like Juniper Hill, virtually nobody knew what it was. I'm willing to bet that a hundred years ago everybody would have known. Back then, a lot of people, maybe even most of them, would have been doing what I do: going out into the fields, woods, moors and coasts looking for wild food to eat. Fifty years ago there would still have been quite a few, but that's about the time when, because of increasing affluence and changing lifestyles, gathering and collecting wild food began to die out. Now people go to the local shops and pay about two quid for a packet of shortbread made with palm oil – crazy!

Afterword

Just as it took me a long time to appreciate the knowledge that I already had, it has taken years for people to rediscover wild food.

As is often the case, things come full circle. I often make the case that humans started off as hunter-gatherers, eating what they could find, including all sorts of insects and bugs. Then we became farmers and we stopped eating bugs. But now, in the twenty-first-century, bugs are having a real moment again, with lots of chefs and food producers experimenting with them. It's the same with many of the plants that are around us. Neither we nor our pigs eat acorns any more, but who's to say they won't become the next food fashion. It's not called the circle of life for nothing.

Writing this book has been an exciting way to pass on this rediscovered knowledge to other people. People don't always realise what it has taken for my family to get to this point. It wasn't just my Grandad's experiences and his earlier life, it was also the learning from our own childhoods. Writing has also reminded me of the amount of research I have had to do, how I've built on my family legacy and got to a point where I am confident in my knowledge and ability as a forager.

It's also reminded me of the times I've been stranded on barbed wire in the middle of nowhere with an irate ram for company, or have forgotten to tie my hair up and have spent half an hour disentangling myself from a hawthorn bush. And the times I've ruined my clothes with burrs or a stray nettle has stung the unmentionables. It is these moments, and many more trials and tribulations, that are absolutely necessary to go through to become the fully fledged forager. I know I have earned my right to do what I do.

I hope I have given people a little glimpse into the weird and wonderful world of foraging. After all I am a woman who picks weeds for a living, and in my world, money *does* grow on trees.

Acknowledgements

To the ones I drive absolutely nuts: my mum Barbara and my husband Chris. Together they are my rock and my anchor and without them I wouldn't be doing what I'm doing. I love them dearly and thank you

To the rest of my family, who have been supportive all the way, and enjoyed my journey with me, especially young Ryan, because he will pick blackberries for me in the rain.

To my wonderful friend and PR guru Bernice Saltzer, without whom none of this would have been possible. Our chance meeting has blossomed into a great friendship. To Neil Hanson who made it all happen, thank you. To Mark Lucas at The Soho Agency who is a legend; I am so glad that he took me on. To my publisher Sarah Emsley at Headline for letting my true voice be heard, and Emma Tait for the sheer effort she has put in.

And finally to all the people I work with in this crazy hospitality industry, which welcomes anyone and everyone. You work like there is no tomorrow; there has never been a group of people who are as generous, kind hearted and gossipy. Thank you to each and every one of you.

Index

Page numbers in **bold** refer to illustrations.

absinthe, 232–3
acorns, 277–80, **279**
Adams, Fanny, 149
Alexanders, 92
allergies, 167, 234
almond trees, 67
Ang (teaching assistant), 34
Angel, The, Hetton, 142
angelica, 218, 219
animal tracks, 184
anti-bacterials, 101
anti-oxidants, 101
Apennines, the, 53
Aperol, 233
appearance, 154
April, 88, 94, 98, 196
arse smart, 134–5
Asda, 59
ash, 7
ash keys, 7, 41
Aspergillus tubingensis, 168
asthma, 230
Atchinson, Craig, 72–3, 142

autumn, 91–2, 93, 205, 266
autumn truffles, 52

bacteria, 167
barren strawberry, 246
Beaumont, Ian, 24
beech, 6
beech nuts, 6–7
bees, 161
bell heather, 22, 129
Berliner Weisse, 235
berries, poisonous, 120
berry season, 91
beta-carotene, 248
Bettaveg, 68
bilberries, 249–51, 264
bilberry picking, 3–5
biodiversity, loss of, 159–63, 163
Birch Syrup Company, 192
birch tree sap, 192–3
Birchall, Mark, 70–1, 79, 99, 127, 196–7, 216
bison grass, 236, 239
bistort pudding, 231
blackberries, 266
blackcurrant leaves, 89

Blackpool, 20
bluebells, 21–2
Bo (dog), 280
boletus, 10
Bolton Abbey, Burlington, 243
Booths, 79
borage, 89
Bosnia, 27
botanical societies, 136–7
bracken, 116–7
Brammer, Barbara (née Szperka),
    3, 12–3, 14, 16, 18–9, 33, 34, 50,
    158, 249–50, 251, 281
  birth, 46
  cooking, 16–7
  finds sweet cicely, 228–9
  as forager, 83–4
  juniper search, 293–4
  moves to Cornwall, 30
  truffle foraging, 55
  truffle growing, 60
  whisky distillery visit, 154–5
Brooke-Taylor, Alisdair, 195
broom, 199
bullace, 264
burdock, 212–4
Burlington, Bolton Abbey, 243
Byrne, Aiden, 69–70, 108, 221,
    269–70

Calderdale Youth Offending Team,
    34
calvacin, 133–4
Camborne College, 31
Campari, 233
cancer treatment, 133–4

cancers, 101
carrots, 216–8
Cartmel, L'Enclume, 70, 127
Cass, George, 27–8
caster sugar, 75–6
cattle, 291–2
ceps mushrooms, 57–8
Charles II, King, 223
chefs, 121, 124–31, 156–7, 175–6
  concerns, 125–6
  enthusiasm, 129–30
  information needs, 127–8
  Michelin stars, 124, 126–7
  needs, 111
  right-hand people, 127
  skills, 125, 126
  supplying, 127–9
  working with, 127–31
chemical properties, 136
cherries, 253–4
Chester, University of, 75
chestnuts, 279
  preparation, 272
  sweet, 270–3, **270**
chicken jalfrezi, 139
chicken of the woods, 239–41, **240**
chickweed, 282–3
childhood diseases, 20
Chinese traditional medicine, 101
Christmas decorations, 8
Christmas Tree Syrup, 72–6, 287
cleaning, 111–2
cleavers, 193–4
Cleethorpes, 85, 100
climate change, 163
clothing, 154

coastal foraging, 27–9, 94–109
  absorption, 103
  and anglers, 105–6
  characters, 104–5
  dangers, 103–4
  mudflats, 103–4
  protected species, 106–7
  range of delights, 95
  seaweed, 107–9
  sewage alerts, 95
  species, 96–102
  summer, 98–102
  tides, 95, 103–4, 111
  weather, 106
Coca-Cola, 222
cockle picking, 103
coeliac, 278
common earthball, 61
compasses, 148–9
conifers, 285–8
Conisbrough Castle, 199
conkers, 8–9, 82
conservation, 102, 159–69
contamination, risk of, 114–5
Copenhagen, Noma, 77–8, 79–81
Cornish Yarg, 186
Cornwall, 26–30, 238
coumarin, 236, 237–8, 238–9
crab apples, 257
Culdrose, RNAS, 26, 27
cultivation, 121–2

Daedalus, HMS, 26
damsons, 258
dandelion, 207–9, **208**
Dandelion and Burdock pop, 212

dead-nettles, 188
death caps, 118–9
December, 288–9
dehydrating, 236
dehydrators, 215
Department for Environment,
    Food and Rural Affairs,
    144–5, 185
destroying angel, 119
devil's boletus, 145
Devon, 29, 142
dewberries, 266–7
diabetics, sugar substitute, 230
dock leaves, 230–2
dock pudding, 231–2
dog mercury, 161
Doncaster, 60, 61, 66–7, 89, 162,
    285
Doogle (dog), 4
Dora, Aunty, 18, 33, 171–2
Douglas-fir-infused butter, 130
Dr Pepper, 222
dryad's saddles, 198–9
dry-stone walls, 163
dusk, 150

East Anglia, University of, 32–3
elder, 203–5, **204**
elderberries, 205
elderflower capers, 204–5
elderflowers, 203–4
endangered and protected species,
    144–5, 185
English summer truffles, 52
environmental consciousness,
    159–69

equipment, 153-4, 200
ersatz coffee, 278
essential oils, 140-1
European larch cones, 77-8
Evans, Nicholas, 119

fat hen, 134
Feeney, John, 131
fermentation studios, 136
fiddlestick heads, 116
field maples, 190-1
finding plants, 146-8
fine dining, 24
flavour, best possible, 121
flavours, 126
    surprising, 135
flies, 87
floods and flooding, 162-3
flour, 278-9, 287-8
flowers, edible, 89
fly-tippers, 163
folk medicine, 101, 195, 230
folklore, 132-3, 282
food intolerances, 234
food poverty, 123
food safety, 206
food safety and hygiene certificate,
    114
Food Safety and Hygiene for
    Catering, 166
food safety and hygiene
    requirements, 73-6
Food Standards Agency, 73-4
food testing labs, 75
foraging, 297
foraging encounters, 150-2

foraging year, the, 86-93
foreign invaders, 144
forest schools, 165-6
Forest Side, Lake District, 243
Four Lanes, 30, 31, 33
Fred (dog), 83, 90, 120, 150, 151-3,
    262
French, Luke, 131, 211
French, Manchester, 232
frogspawn, 6, 163
frozen wood ant bum, 207
fruit season, 91
fruits, picking, 115

Gallows Pole Hill, 4
gardens, loss of, 162
geese, 21
Genea truffles, 55, 56-7
genetic modification, 122-3
giant hogweed, 218
Gidleigh Park, Devon, 142
gins, 138-40, 141-2, 215, 219, 226-7,
    264-5, 291
girolles, 243-5, **243**
Glen (friend), 23
gloves, 264, 276
gobo root, 213
golden plovers, 22
golden truffle fly, 54
Goldstraw, Aaron, 78-80
goosegrass, 21
gorse, 199-201, **200**
gorse flowers, 87, 199, 200-1
Gosport, 26
GPS details, 147
Grand Hotel, York, 142

green walnuts, 275, 276-7
grouse, 129-30
growing conditions, 121-3

Halifax, 31
hares, 22, 161-2
Harome, Star Inn, 98, 99, 127, 277
Haworth, Nigel, 80
hawthorn, 256-7
hawthorn flowers, 87
hay-baking, 236
hazelnuts, 275-6
heartsease, 114
heather, 22, 128-9
heather pollen, 129
hedgehog fungus, 145
Helston, 26
hemlock water dropwort, 217-8, 228
herons, 89-90
Herstell, Chris, 68
Hetton, Angel, The, 142
Himmler, Heinrich, 37
Hipping Hall, 233
hogweed, 137, 218
holly, 8, 147
Holocaust, the, 38
honey, 49-51
Hope & Anchor, north Lincolnshire, 72, 85-6
horse chestnuts, 8-9
horse flies, 116
horseradish, 209-11, **210**
Huddersfield, 17, 231
Humber estuary, 95, 96
hyper-vigilance, 112

identification, importance of correct, 116-21
immune system, supercharged, 20
ingredients., fresh, 121
insects, 87, 115-6, 160-1, 292-3
Institute of Food Science and Technology, 166
insurance, 75
Irish strawberry tree, 247-8

Jack-by-the-hedge, 88
Jack-in-the-pulpit, 65, 183
James Martin's Saturday Morning (TV show), 227, 261, 276
Japanese knotweed, 144
Japanese rosehips, 263
Jews' ears, 86, 194-6
Jöro, Sheffield, 131, 211
July, 89
June, 88, 99
juniper, 139, 140-2, 288-95
    appearance, 289-90, **289**
    berries, 290-1
    spiders, 292

Karen (food technologist), 73-4
Kew, Royal Botanic Gardens, 54
Killarney tree, 247-8
knives, 153-4
knowledge, 146
    passing on, 176, 297
    rediscovered, 297
    sources of, 132

Ladstone Rock, 4
Lake District, Forest Side, 243

Lakeland, 73-6
Lancashire, Moor Hall, 70-1, 79, 99
Le Cochon Aveugle, York, 276
lead, 114
leaves, patterns, 147
Leeds, Royal Naval Recruitment
    Centre, 25
L'Enclume, Cartmel, 70, 127
Leonard, Paul, 243
lichens, 283, 285
Lime leaves, 89
Lincolnshire coast, 95
Lindsay (chef), 24
lingonberries, 84, 137, 251-3, 252
liquorice root, 59
Little Brown Mushrooms, 118
Little Harrow, the, 57
liver flukes, 206
London, 33
lords and ladies, 65, 183
Lorraine (aunt), 23
Louis XIV, King of France, 223
Lovatt, James, 127
low ovens, 215

Macedonia, 77-8
McGurran, Colin, 71-2
Malham Cove, 206-7
Manchester, 69
    French, 232
    Midland Hotel, 70
    New Smithfield Market, 67-8
mangetouts, 24
March, 87-8, 192
Marilyn (Aunt), 20, 47, 157, 171
Martin, Oli, 233-4

Martini, 233
May, 88, 94, 98, 201
mayweed, 135
Maywein, 237
mead wine, 237-8
meadowsweet, **235**, 236-9
medicinal plants, 133-4, 195, 261-2
medlars, 267-70, 268
melilot, 236
midges, 115-6
Midland Hotel, Manchester, 70
Midsummer's Day, 99
Mikolajczyk, Slawomir, 71-2, 85
milk-cap mushrooms, 242
mindful foraging, 144
Moor Hall, Lancashire, 70-1, 79,
    99, 127, 196-7, 216
Moorcock Inn, Norland, 195
Morecambe Bay, 103
Morello cherry trees, 67
morels, 119-20, 201-2, **202**
mosquitoes, 88, 116
moths, 10
motorways, 113
Mr P's Tavern, York, 186
muck, 20
mugwort, 233-4
Mushroom Martin, 63-5, 66, 67,
    69, 116-7, 218
mushrooms, 10, 112-3, 146-7
    availability, 88
    boletus, 10
    ceps, 57-8
    chicken of the woods, 239-41,
        240
    cleaning, 196, 201

common earthball, 61
death caps, 118-9
destroying angel, 119
dryad's saddles, 198-9
drying, 195-6
edible, 117
giant puffballs, 61-4, 88, 133-4,
   168
girolles, 243-5, **243**
identification, 117-20
Jews' ears, 86, 194-6
LBMs, 118
lion's mane, 242
milk-cap mushrooms, 242
morels, 119-20, 201-2, **202**
oyster mushrooms, 241
panther cap, 117
parasol, 117
penny buns, 57-8
poisonous, 117-20, 240-1
porcini, 57-8
scarlet elf cups, 118
St George's, 40, 196-8, **197**
myrrh gum, 141

name dropping, 77
National Parks, 162
nature, 10
navigation, 147-9
nettles, 186-8, **187**
   preparation, 187-8
   stings, 188, 230-1
New Covent Garden Market, 67
New Smithfield Market,
   Manchester, 67-8
Newquay, 30

newts, 163
Noma, Copenhagen, 77-8, 79-81
Norfolk, 63-4, 67
Norland, 54
   Moorcock Inn, 195
Norland Moor, 3-4
Norway, 32
Nouvelle Cuisine, 24
November, 88, 283
nuisance species, 29
nut season, 91-2
nuts, picking, 115

oak trees, 91
O'Keeffe, Jess, 226
Oliver Kay, 78-80
organic foraged food, 68
Organic North, 68
orris root, 215
Over The Bridge, Ripponden, 24
Overington, Josh, 276
overpopulation, 123
oyster mushrooms, 241

P&O, 32, 33
packaging, 111
Padstow (dog), 27-8, 31, 33
panther cap, 117
parasol mushrooms, 117
parsnips, 216-8
Paul (friend), 61
Peak District, 290
pectin, 257
Pendeen, 27-8, 238
penny bun mushrooms, 57
pennywort, 28-9

perfume, 140-1, 222-4
permissions, 144
Pern, Andrew, 98, 99, 127, 186
pesticides, 160, 161
pheasant, 24
phloem, 287-8
picking, timing, 115-6
pignuts, 219-21, **220**
pine martens, 74
pine needles, 74
pine nuts, 287
pine trees, 287-8
pink larch cones, 79-81
plant lore, 132-3
plants, common names, 132,
    134-5
plastics, 167-8
plums, 257-60, 264
Plymouth, 25-6
poisonings, 61
poisons, 114, 217-8, 240-1
Poland, 35-47, 71-2, 117-8
Police Wives' Recipe Book, The, 17
pollution, 90, 114, 159
pomanders, 222-4
Pontefract, 59
porcini mushrooms, 57-8
Portheras Cove, 28-9
Poznań, 37-9
preserving, 230
processed food, 121
protected species, 106-7
publicity, 79
puffballs, 61-4, 88, 133-4, 168

quality, commitment to, 112-5

R and R, 89-91
Radjel Inn, Pendeen, 27-8
Raleigh, HMS, 26
red grouse, 22
Redzepi, René, 77-8, 79-81, 275
regular spots, 147
Reid, Adam, 70, 232
reindeer moss, 283-5, **284**
reliability, 81
reputation, importance of, 76-7
research, 297
restaurants, 68-72
rhubarb, 189
Ripponden, 12-22, 49, 162
    Over The Bridge, 24
River Don, 90
River Ryburn, 159
roadside rhubarb, 92
Robert (uncle), 3
rock samphire, 28
Rogan, Simon, 70, 127
root beer, 226
roots, 209, 211
rosehips, 260-4
    Japanese, 263
roseroot, 207
rowan, 254-6, 255
Royal Botanic Gardens, Kew, 54
Royal Navy, 25-30, 31, 149
Ryburn Valley, 12

Safe and Local Supplier Approval,
    166
saffron milk-cap, 243-4
St George's mushrooms, 40, 196-8,
    197

salmonella poisoning, 114
salt marshes, 102–4
samphire, 98–100, 103, 106
sand-worms, 105–6
SAS-type survival techniques, 228
Savernake Forest, 57
scarlet elf cups, 118
Scotland, 189, 288–9, 290, 293–4
scurvy grass, 96
sea anemones, 28
sea arrowgrass, 98
sea asparagus, 97
sea aster, 97–8
sea beet, 97
sea blite, 96
sea buckthorn, 100–2
sea coriander, 98
sea herbs, 95
sea kale, 106–7
sea kelp, 107–8
Sea King helicopters, 27
sea purslane, 96, 102
sea rosemary, 96
searching, 146–8
seaweed, 107–9
Second World War, 36–47
Sekulka, Mr, 173–4
September, 99
sewage alerts, 114–5
Sheffield, Jöro, 131, 211
shellfish, 104–5
shepherds, 169
Shit with sugar on it, 17
silverweed, 281–2
sloes, 264–6, 265
Smith, Stephen, 127, 277

Soil Association, 68
Southport Food Bank, 78–9
Sowerby Bridge, 84, 162, 170
spare time, 110
Speedo man, 150
spiders, 116
spindle trees, 120
spring, 86–7, 115
squirrels, 74, 91
Star Inn, Harome, 98, 99, 127, 277
Star Inn The Harbour, Whitby, 98
Stein, Rick, 28
stonecrop, 206–7
storage, 137
storages, 111–2
strawberries, 234, 244–7, **246**
Sue (chef), 24
summer, 115
sustainability, 282–3, 290
Sutcliffe, Adrian, 3, 12, 13, 15–6, 19–20, 31, 84, 251
Sutcliffe, Alex, 84, 176
Sutcliffe, Ryan, 84–6, 176, 188, 221, 249, 281
swallows, 93
sweet chestnuts, 9, 270–3, **270**
sweet cicely, 136, 227–30, **229**
sweet flag, 221–7, **224**
    citrus scent and taste, 221
    first find, 224–5
    imported, 222
    introduction to Britain, 222
    processing, 226–7
    root, 225
    use as a perfume, 222–4

sweet woodruff, 136, 234–6, **235**, 239

syphilis, 223

Szperka, Dan (Bogdan Adam Stefan) (grandfather), 169–73
  arrest, 39
  and cement, 172
  escape, 39
  family background, 35–6
  frailness, 172–3
  harmonica, 37
  and invasion of Poland, 36–7
  leaves Poland, 45
  marriage, 46
  obsession with food, 170–1
  with the Polish Resistance, 43–5
  at Poznań, 37–9
  refuses medals, 169–70
  on the run, 39–43
  story, 35–47
  teatime with, 5–6
  threat of collaborators, 42–3
  walks with, 3–5, 6–8, 10, 11, 173–5, 249, 270–3

Szperka, Isydor, 35, 36, 38–9, 46–7

Szperka, Magdalene, 35, 46–7

Szperka, Robert, 170, 171

Szperka, Ted (Tadeusz), 35, 36, 38–9, 39–45, 47

Szperka, Winnie (grandmother), 45–6, 50, 170, 171–2

tangerine root, 221

tansy beetles, 160

taste
  enhancing, 137–8
  science of, 135–8

Taxotere, 133

terroir, 113

Thatcher, Margaret, 14

Thomas, Dr Paul, 58

thrift, 28

thujone, 232–3

tides, 95, 103–4, 111

Tiggy (dog), 31, 33, 49–50, 54

The Times, Tomorrow's Lawyer competition, 32–3

timing, 115–6, 149

Tom the Cockle Man, 104–5

tonka beans, 236, 239

Torpoint, 26

Toulouse-Lautrec, Henri, 232–3

toxins, 112–4, 120

Trewellard Meadery, 238

truffles, 85
  appearance, 52
  aroma, 52, 54
  autumn truffles, 52
  commercial, 56–7
  continental, 52
  English summer truffles, 52
  flesh, 52
  foraging for, 53–8
  forgery, 52–3
  Genea truffles, 55, 56–7
  growing, 58–60
  hunting, 51–2
  mycelia, 54
  prices, 52
  symbiotic relationship with trees, 54

Turkish delight, 261

20 Stories, Manchester, 108
Twitter, 77, 80

Vasey, Alysia
  attachment to Yorkshire, 169–77
  becomes professional forager,
      60–5
  begins foraging career, 49–65
  broken hand, 93
  childhood, 12–22
  childhood diet, 20
  death of father, 12–3
  develops customer base, 67–76
  early cooking experience, 16–7
  early experiences of death, 23
  endorsement from René
      Redzepi, 80–1
  engagement at nineteen, 25
  enlists in Royal Navy, 25
  family background, 35–6
  father, 12, 16
  find first truffle, 54–5
  first experience of fine dining,
      24
  first jobs, 23–4
  first Michelin-starred customer,
      70
  foraging encounters, 150–2
  grandfather's story, 35–47
  introduction to foraging, 3–11
  learning curve, 76
  leaves Royal Navy, 31
  loses wellies, 18–9
  meets Chris, 48
  mission, 66
  move back to Yorkshire, 33

  passing on knowledge, 176, 297
  Polish roots, 5–6
  railway line woodland walks,
      15–6
  relationship with foraging,
      143–6, 175–7
  reputation, 76–80
  Royal Navy career, 25–30
  at school, 8–10
  as teaching assistant, 34–5
  teatime at Grandad, 5–6
  television appearances, 82
  truffle foraging, 53–8
  truffle growing, 58–60
  truffle obsession, 53, 58
  UN peacekeeping service, 27
  university education, 31–3
  walks with Grandad, 3–5, 6–8,
      10, 11, 173–5, 249, 270–3
  wasp stings, 49–51
  wedding, 48–9
  weekends with grandparents,
      17–8
  whisky distillery visit, 154–5
Vasey, Chris, 64–5, 150, 158, 189, 191,
      244–5
  and customer interaction, 83
  and fine dining, 79, 83
  as forager, 82–3, 111
  juniper search, 293–4
  meets Alysia, 48–9
  redundancy, 66
  rescued by a bunch of kids,
      258–9
  wedding, 48–9
vinaigrettes, 235

vitamin C, 134, 248, 256, 260-1, 282, 287

Waldmeister, 237
Wales, 201
walnuts, 9-10, 273-5, 276-7
Warsaw Uprising, 44
wasps, 49-51
water pepper, 134-5
water plants, 205-6
water run-off, 162
watercress, 205-6
waxworms, 168
weather, 86, 88, 92-3, 106
Wessex helicopters, 26
wheat, 278-9
whisky chips, 155-8
whisky distilleries, 154-5
Whitebridge, 293-4
wholesalers, 67
wholesalers' etiquette, 67-8
Wignall, Michael, 142
wild garlic, 64-5, 86, 152, 181-5, **182**
    benefits, 182-3
    dangers, 183
    decomposition, 185
    growing season, 181
    picking, 182, 183-4
wild garlic capers, 69-70, 184-5

wild plants, survival, 121-3
Wildlife and Countryside Act, 106
Wildlife Trust, 164
winter, 92
Winter of Discontent, 13-4
Winteringham Fields, north
    Lincolnshire, 71-2
winterweed, 282-3
Witkowski, Tadeus, 43
wood and water avens, 214-5, **214**
wood frogs, 65
wood pigeons, 4-5
wood sorrel, 189-90
woodlice, 201
wormwood, 232-3
wrecking, 30-1

yew platters, 285-6, 287
yew tree, 132-3, 133, 240-1, 285-7
York
    Grand Hotel, 142
    Le Cochon Aveugle, 276
    Mr P's Tavern, 186
Yorkshire Forager Gin, 142
Yorkshire Foragers
    develops customer base, 67-76
    first Michelin-starred customer,
        70
    foundation, 66